saturn

About The Author

Jakob Lorber was born on July 22, 1800, on the left bank of the River Drau amidst vineyards, in the Village of Kanischa and parish of Jahring, where his father, Michael Lorber, labored on a small farm.

It was not mere coincidence that Jakob Lorber grew up in an impoverished rural environment, though in a home open to art and religion. He inherited from his father his many-sided musical talents, receiving instruction on the violin, piano, and organ.

By the time he attended the gymnasium (high school) in Marburg, Lorber had earned the necessary tuition money as an organist at one of the local churches. He received his accreditation as a high school teacher in 1829, in Graz, Austria, capital of the province of Steiermark. At that time, however, he could not find appropriate employment. This prompted him to continue his musical studies, which consisted of composing, teaching the violin, providing singing lesson, and giving an occasional concert.

During these years, Jakob Lorber followed his inclination and immersed himself more deeply in the spiritual "Path to the Innermost." He read, among others, the writings of Justinus Kerner, Jung-Stilling, Swedenborg, Jakob Böhme, and Johann Tennhardt. The Bible, however, was his constant companion and remained his principal source of inspiration until the end of his life.

Despite his many abilities, he lived from hand to mouth, until he was finally offered a position as conductor of the Opera in Trieste. It was just as he was about to accept this position that he received his appointment as "God's scribe." On March 15, 1840, right after early morning prayer, he heard a voice in his heart which very clearly ordered him: "Get up, take your pen and write!"

He abandoned all travel preparations, and obediently sat down to write what the mysterious voice dictated. It proved to be the introduction to his first work, The Household of God (Die Haushaltung Gottes): "And thusly the good Lord speaks to everyone; and this is true, faithful, and certain. Whosoever wishes to speak to Me should come to Me, and I will place the answer in his heart. But only the pure, whose hearts are full of humility, will hear the sound of My voice. And whosoever prefers Me to all things and to the world, and loves Me like a bride loves her groom, with such a person I will walk arm in arm. Such a human being will for all times look upon Me as one brother would look upon another, which is how I have looked upon him from eternity, before he existed."

From the hour of this first dictation by the Lord, the unexpected and unheard powerfully entered Jakob Lorber's life. During the twenty-four years that followed, he wrote almost daily for many hours without interruption, neither consulting any books of reference nor otherwise having any prior experience of the voluminous knowledge which flowed from his pen through the Inner Word. His life was fulfilled solely in obedience to this Inner Voice.

One would have to speak entirely in superlatives in order to capture the essence of Jakob Lorber. If we consider him as a literary man, then he surpasses all other authors, poets, and thinkers throughout history. Where else can one find such comprehensive wisdom — interpretations of such depths, or such accurate knowledge of geography, history, biology and natural science, even unto the creation of the cosmos? His major works fill twenty-five volumes of 500 pages each, not counting his other smaller volumes. And if we consider him as a mystical genius, he surpasses all other initiates of whom we have knowledge. No words can encompass him; and if he called himself "God's scribe," that was only his own humble self-assessment.

Jakob Lorber died on August 24, 1864. His mission was accomplished, and he had foreseen his own death. On his tombstone at St. Leonhard Cemetery in Graz, the apostle Paul's words are written: "Whether we live therefore, or die, we are the Lord's.

saturn

by jakob lorber

a description of saturn
the ring, moons and its living creatures

merkur publishing inc.
wisdom of the occident

First English Edition 1998
ISBN 1-885928-07-6

Translated by Gerhard Hanswille & Franca Gallo from the German version *Der Saturn* (1969) ISBN 3-87495-048-4

Printed in the United States of America

Cover design:
Reed Perkins
1229 Buccaneer Drive
Salt Lake City, UT
84117
(801) 596-7084

table of contents

chapter 1 17
The Actual Name And Description Of Saturn
The Double Ring And Moons Of Saturn
The Magnificence Of Divine Revelation

chapter 2 22
The Water-Rich Surface Of Saturn
The Continental Islands
Great Glacial Zones At The Poles
The Mild And Pure Central Zone
Intense Light Atmosphere, Light
And Temperature Conditions
Saturn's Ring As A Regulator
And Reflector
The Splendor Of the Starry Sky

chapter 3 25
The Land Herrifa
The Healing Mountain, *Girp*
The Sun Tree, The Rain Tree
The Hair Tree, The Broad Tree
And The Ray Tree

chapter 4 33
The Funnel Tree
The Pyramid Tree
The Mirror Tree

chapter 5 39
The Hotchpotch Tree
The Fire Tree And Oil Bush
The Physical Form Of
Human Beings On Saturn
Their Real Property And Domesticated
Animals

chapter 6 45
The Herbs And The Useful Plants Of Saturn
Corn
The Flowing Barrel And
The Walking Bottle Gourd

chapter 7 52
Saturn's Wealth In The Vegetable Kingdom
Saturn's Primary Color, Blue
Aromatic Healing Herbs
The Gold Shrub, Metal Plants, Blue Grass
Form-Changing Meadow Flowers,
Fragrant Algae Moss
Saturn's Mountains And Plains

chapter 8 60
The Ship Plant, *Chaiaba*

chapter 9 65
Additional Information About The Ship Plant
The Streams Of Saturn And Their Even Descent
The Sphere Shell Structure Of Saturn
And All The Other Celestial Bodies
The Wealth Of Forms Of Creation
The Four Main Streams Of Saturn

chapter 10 72
The People Living Adjacent To
The Morning Stream
The Tree Dwellings Of Saturnites
How The Will-Power Of The Saturnites
Controls Nature
Saturnite Communication With
The Spiritual World
The Useless Missions Of
Earth Spirits To Saturn

chapter 11 78
The Beautiful Riparian Landscapes
The North Stream
The Evening Stream
The Midday Stream

chapter 12 81
The Inland Lakes On Saturn
The Magnificent Countryside
And The Large Population At The Shores
Links Between Rivers, Streams And Lakes
Stone Cone Groups As
Places Of Entertainment
Swan Rides

chapter 13 89
The Shores Of The Oceans
Danger Through Storm Floods
The Moons And The Rings As
The Originators Of Floods
The Lower Species Of Ocean Animals
The Giant Blue Mussel

chapter 14 95
The Stick Snail
The Pyramid Snail
The Disk Snail That Produces
Ointments And Garden Ornaments
For The Patriarchs' Coats

chapter 15 105
The Seven Snail
Practical Utilization Of Its Shell
Weight Ratios On Saturn

chapter 16 109
The Giant, Illuminated Ray Snail
Its Appearance During Ocean Storms

chapter 17 111
More Information About Water Animals
The Largest Whale-Like Fish — *Bisorhiohiohio*
A Transformer Of the Spiritual Astral
Of Water Animals To Animals Of The Air

chapter 18 115
Flying Insects
The Saturn Fly
The Flying Star
The Giant Butterfly *Com*
And The Utilization Of Its Plumage

chapter 19 120
The Different Kinds Of Bats
The Flying Cow
The Flying Ribbon
The Jewelry Trade On Saturn

chapter 20 125
The Host Of Flying Animals
Without Plumage
The Kingdom Of Birds
The Water Hen *Behor:* A Large
Species Of Heron
Heaven's Messenger
The Flight-Tone Accord Of These Birds
Song And Music Of The Saturnites

chapter 21 131
The Singers Above The Rivers And Lakes
The Masters Of The Art Of The Fugue
The Northern Singer Of The Air,
The Most Effective Music

chapter 22 138
The Domestic Hen
The Golden Sphere
The Giant Goose
The Nature And Benefits Of
These Domesticated Birds

chapter 23 144
The Largest And Most Important Land Animal
—The *Mud*
The Teachings Of The Wise About The *Mud*
The Purpose Of Its Creation

chapter 24 149
The Saturnian Elephant, *Sisterkihi*
Its Structure And Habits
Hunting This Animal And Its Purpose Of Creation

chapter 25 153
The Blue Bear, *Ihur*
Its Structure, Character And Nourishment
Its Usefulness As A Cultivator Of The Soil

chapter 26 157
The Saturnian Lion, *Horud*, Serves As
A Hunter And Lumberjack
Method Of Catching The Young
Saturnian Lions

chapter 27 162
The Saturnian Antelope, *Zigst*
Its Purpose And Creation Hints
The Reasons For Hunting *Zigst*
Fraud With Secret Remedy

chapter 28 169
The One-Eyed *Bauor*
Hunting The Animal
The Skin As A Coat For The Patriarchs

chapter 29 174
Harmony Of The Celestial Bodies
Examples Of The Art Of Music
The Secrets Of The Teachings
Of Sound And Creation
Correspondence Of The Wild Animals
On Saturn And On Earth

chapter 30 181
The Tame Animals Of Saturn
The Most Beneficial Animal,
The Cow, *Buka*

chapter 31 185
The Blue Goat
Bartering With Its Milk
The Goat Thanksgiving Festival
The Connection Of The Saturnites
With The Spiritual World
The Wool Of The Blue Goat And Its Use

chapter 32 190
The Saturn House Servant, *Fur* –
A Species Of Tame Ape
Concerning The Saturn Dog, Horse, Sheep
And Other Animal Species

chapter 33 195
The Saturnian Human Beings
Their Origin, Population Density And
Housing Conditions

chapter 34 203
The Household Constitution Of The Saturnites
The Planting Of A Temple
The Most Important Law: Emigration
Love Thy Neighbor Or Charity

chapter 35 213
The Main Law Of Life: The Will Of God
Treatment For Offenders

chapter 36 218
The Metal Industry And Other Crafts Or Trades
True Socialism Of Altruism In
A Trade And Commerce

chapter 37 ... 223
The Manufacture Of Woven Material
No Unnatural Utilization Of Natural Products
The Clothing Order

chapter 38 ... 226
The Stature Of A Man And A Woman
Procreation, Pregnancy And Birth
The Constitution Of The Children

chapter 39 ... 231
The Cognizance Of God By The Saturnites
More Respect Than Love
Knowledge Of God's Incarnation On Earth

chapter 40 ... 233
Law Against Arrogance
The Law Of Cleanliness
Aversion To Everything Dead
Funeral Services
Marriage

chapter 41 ... 239
The Inhabitants Of The Lowlands Visit The Mountains
Their Reception By The Mountain Dwellers
Half-Heathens And Their Conversion
Courting In The Lowlands

chapter 42 ... 246
The Innermost Spiritual Religion Of The Saturnites
The Significance Of The Number Seven
Sunday Celebration
The Baptism Or Blessing Of A Newborn Child
The Temple Meal
The Sermon Of The Elder
Supported By Spiritual Visualization
The Wisdom Of The Saturnites

chapter 43 250
Profound Spiritual Observations Of Nature
Communication With Angels And The Lord Himself

chapter 44 253
The Spiritual Aspect Of The Saturnian Religion
Spiritual Rebirth Through Teachings And Practice

chapter 45 257
More Details Regarding The Path Of Rebirth
And Becoming One With God

chapter 46 260
The Inhabitants Of The Lowlands
Their Construction And Housing
And Their More Worldly Way Of Thinking

chapter 47 263
More About The Inhabitants Of The Lowlands
Commerce And Trade
Banishment Of Dishonest People
Mode Of Life, Matrimony, Procreation And Burial

chapter 48 267
The Vast Northern And Southern Polar Regions
The Enormous Coldness Serves To Temper The Destructive
Impulse Of The Primary Spirits Of This Planet
How Saturn Affects The Earth

chapter 49 271
Saturn's Ring Has A Planetary Character
Axis Rotation And Its Intended Purpose

chapter 50 274
More About The Various Purposes
Of Saturn's Ring
An Example Of Glass-Blowing And
The Cooling Oven
The Delicately Material Ring Inhabitants
Their Way Of Life And Their Communication
With The Inhabitants Of The Planet
Absence Of The Animal World And
Seedless Fruit Trees

chapter 51 278
The Saturn Moons Without Any Axis Rotation
Natural Life Exists Only On The Opposite
Side Of The Moon
The Moons As Post-Schools Of Saturn
And Pre-Schools Of The Ring And
The Purely Spiritual Life
Closing Words

1

The Actual Name And Description Of Saturn
The Double Ring And Moons Of Saturn
The Magnificence Of Divine Revelation

1. In order to have a clear concept of this planetary body which you call Saturn, it is important to know what its actual name denotes: Earth Calmness, World Nothingdom. It is also absolutely necessary to learn about its natural sphere, its distance from the sun, its size, structure, its inhabitants as well as the inhabitants on the rings and moons, as well as its diverse vegetation in accordance with the conditions that prevail because of distinctly varying climates. Furthermore a description of all the animals that exist on this planet, its rings and moons.

2. Once all the above-mentioned details about Saturn have been properly explained will the history of this planet, its inner structure and polarity to other planets, and finally the spiritual sphere, be analyzed.

3. As far as the distance of Saturn from the sun is concerned, three different points of view can be adopted; the reason for this is well known. There is not one planet whose orbit is on a completely circular course; instead, a planetary orbit around the sun is like an ellipse whereby the sun is in relationship to the orbit of a planet, similar to an egg with the obtuse part downwards and the acute part facing upwards. In this position the egg yolk will not be in the center of the egg; instead it will be considerably closer to the bottom, the obtuse part. Let us assume that the central egg yolk represents the sun and the periphery of the eggshell the orbit of the planet. If you measure the distance from the periphery to the center of the sun, in this instance the egg yolk, you will have the following

results: The center of the sun is closest to the periphery at the lowest part. At the girth the distance is at a middle distance, whereas at the upper tip the distance is the greatest from the center of the sun. The same applies to the orbit of Saturn around the sun. When Saturn is at the lowest point, the distance to the sun is only 187,719,120 geographic miles.[1] When Saturn's orbit is at about the girth, the distance is already 198,984,136 geographic miles, and at the top the distance from the center of the sun is 210,249,152 geographic miles, and the latter distance is of course the greatest from the sun.

4. These distances are not measured from the earth, but from the sun. The distance of the earth in relationship to Saturn can vary tremendously, even to the extent that these two celestial bodies could be closer to each other by one million geographic miles, and then again they could be farther apart by one million geographic miles. When it should happen that both these planets are on one and the same side from the sun, then both planets are in the proximity to the sun. In this position they are much closer to each other's proximity than when they are in opposition, where it can occur that Saturn is the farthest from the sun, whereas the earth, on the opposite side, could be closest to the sun. When that occurs the difference is not only one million, but often two to three million, geographic miles. The reason why exact distances cannot be given is that not one planet orbits always in exactly the same distance from the sun; instead, in one year the planet distances itself while in another year it comes closer to the sun, and the greater or lesser distance determines the temperature differences. And you can be assured that, of seventy-seven orbits, not even two orbits as far as the distance from the sun is concerned are completely the same.

---

[1] One geographic mile = 7.420 km; Austrian mile = 7.586 km. – ED.

5. Since we have completed the distances, we will now determine the diameter of Saturn as well as the circumference, its surface in square miles and its content in cubic miles.

6. The diameter of Saturn is 17,263 geographic miles. The earth in comparison is only 1,719 geographic miles; this will give you an idea by how much Saturn is larger. Saturn's circumference is 54,515 geographic miles. The surface is 936,530,820 square geographic miles. The cubic content of Saturn is 2,757,547,946,775 cubic geographic miles. Therefore, Saturn is approximately 1,037 times larger than earth. For one orbit around the sun Saturn requires 29 years, 164 to 166 days, 2 hours and 2 seconds.

7. Everything which can be counted as far as Saturn itself is concerned has been determined. Since Saturn is surrounded by a double ring, this must also be more closely determined in the form of numbers.

8. The diameter of the entire ring is 40,006 geographic miles. Since the ring actually consists of two rings, the distance from the surface of the inner ring to the inner surface of the outer ring is 545 geographic miles. The diameter of the outer ring, that is from the outside to the inside, is 1,350 geographic miles; and the diameter of the inner ring, measured in the same manner, is 3,850 geographic miles, since the outer ring as well as the inner ring are elliptical (oviform); that is, if you cut through the ring it has the shape of an egg. The diameter of the outer ring at the girth is 130 geographic miles. The inner ring has within itself three semi-divides, each of which is 20 to 30 geographic miles. These divides are called semi-divides because they do not divide the entire second ring completely, the way the outer ring is separated from the inner ring. These three semi-divides are filled with nothing but oviform spheres which have a diameter large enough to enable these spheres to form only one

ring. However, there is a space where these divides are; it goes inward like an arched-in pyramid, from below upwards and from above downwards, throughout the entire ring. These strings of spheres in these three semi-divides have caused many keen-sighted astronomers to assume that this ring is composed merely of many moons, because through a telescope it has the appearance of a rosary. But this is not the case; instead, they are merely many small spheres.

9. As far as further details in regards to the structure of the ring is concerned, these will be explained later on. Now we will take a brief look at the moons of this planet.

10. Seven moons of various sizes orbit Saturn; they orbit at different distances from Saturn.[2] The first moon, which is the closest and at the same time the smallest of the moons, is only 120 geographic miles in diameter and the distance from Saturn is 28,840 geographic miles (this distance is the mean distance). The second moon has a diameter of 240 geographic miles and is at a distance of 40,516 geographic miles from Saturn. The third moon has a diameter of 666 geographic miles and is at a distance of 60,500 geographic miles from Saturn. The fourth moon has a diameter of 699 geographic miles and is at a distance of 87,920 geographic miles from Saturn. The fifth moon has a diameter of 764 geographic miles and is at a distance of 190,000 geographic miles from Saturn. The sixth moon has a diameter of 900 geographic miles and is at a distance of 277,880 geographic miles from Saturn. The seventh moon has a diameter of 1,120 geographic miles and is at a distance of 360,920 geographic miles from Saturn.

---

[2] According to the current scientific paradigm there are ten moons orbiting Saturn; however, three of them are actually asteroids. – ED.

11. From the information which you have been given so far, you may conclude that this celestial body, by virtue of its size, its different structure and also by virtue of its seven moons, plays an important role in the realms of creation.

12. Because the more artfully a mechanic constructs his work, the more multifarious must be the purpose of such a work. And just like the mechanic who has incorporated various determinations in an artful work in order to attain various purposes, I, as the Greatest Mechanic in the universe, would not place such a celestial body so artfully into the vastness of space without a great significant purpose. Since I consider even the smallest particle of solar dust to be significant, how much more important must a celestial body such as this great planet be? I did not create it to be a mere toy.

13. As a consequence of this revelation regarding this celestial body, you will be taught its purpose from such a sublime aspect that it will rob you of your breath. You were already surprised and experienced certain emotions when I revealed the moon.[3] But how will you react when you travel to this celestial body with Me for a little while? All I can say is that you will experience great things therefore prepare yourself. You will hardly believe what you will see. Whenever I reveal such great things, the only beings that belong are beings who have a big heart in order to comprehend and honor this greatness. Once you have received all the information about this celestial body in so far as it is possible for you to comprehend, then you will begin to appreciate the meaning of the verse in the Bible which states: Eye hath not seen, nor ear heard, neither have entered into the heart of man, the things which God hath prepared for them that love Him. (No human eye has seen, no human ear has heard,

[3] This information can be found in *Earth & Moon* by Jakob Lorber. – ED.

and never has it entered into a man's heart or mind that which God has prepared for those that love Him.)

14. Whoever receives something from Me will always receive the highest gift of heaven, because I am the Highest of All in heaven, in the universe, and in all worlds. And should I reveal to you heaven or hell, one as well as the other will at all times contribute to your blissfulness. Whatever My word may imply, it is alive throughout and makes the one who receives it and accepts it in all love, gratitude, humility and in living belief, eternally alive and thereby already within Me, here as well as in the beyond, and through this preference extremely blissful.

2

The Water-Rich Surface Of Saturn
The Continental Islands
Great Glacial Zones At The Poles
The Mild And Pure Central Zone
Intense Light Atmosphere, Light And Temperature Conditions
Saturn's Ring As A Regulator And Reflector
The Splendor Of The Starry Sky

1. Since we have been given the required introduction, we can dare and begin to take a look at the planet itself.

2. Observe Saturn's surface: The greatest part is water. On this planet there is no actual continent, but below the equator there are isolated major islands which are larger than Europe, Asia, Africa, America and Australia combined. However, because of Saturn's size they cannot be considered as continents, but rather as islands, and they are farther apart from each other than Asia and America in the

region of the equator on earth. In between these larger islands there are of course quite a number of smaller islands, which in proportion to the larger islands are the same as the smaller islands compared to the continents on earth.

3. Towards the poles this celestial body is covered with permanent ice and snow; this zone begins at 40° (latitude) earlier than it does on earth. And what is described on earth as the moderate temperature zone is a region of snow on Saturn. And what is the frigid zone on earth is the region of permanent ice on Saturn. The zone which, on earth, is tropical is the moderate or pure zone on Saturn, where clouds or fog very seldom form, whereas the two other zones are under permanent fog and cloud.

4. As rough and severe as are the northern and southern snow and ice zones, so bright, mild and pure is the central zone, which is the only inhabitable zone. In this zone there are seventy-seven large islands, of which a medium-sized island is larger than the Americas. And each of these islands is, in its formation as well as in its products, more varied by far than Lapland is from the most southern tropical countries on earth.

5. You will probably think that Saturn, because of its great distance from the sun, would be fairly dark, and that the temperature would not be too warm even at the equator. If you think that way, you are in error, because this planet has its own light, which is proportionately stronger to the same degree as it is larger than earth. This planet is also surrounded by an atmosphere which is a thousand times larger and farther reaching, with a diameter of almost 100,000 geographic miles, whereas the earth's atmosphere does not even measure 2,000 geographic miles in diameter, inclusive of the earth's diameter. Since Saturn's atmosphere has such an extraordinarily large diameter, how many sunrays is this large air

sphere capable of absorbing in order to conduct them in a broken line and in a more condensed form to the surface of this planet? This is also why the inhabitants of Saturn see the sun as being many times larger than do the inhabitants of earth. And through this, the heat around the equator would be unbearable if it were not tempered by the surrounding ring, which absorbs most of the condensed sunrays and partly makes use of them itself; the remainder it returns to the universe. That is why, through a telescope, the ring appears more illuminated than the planet itself, while its shadow has a very beneficial effect upon Saturn and, through this, the tropical or hot zone becomes a moderate zone.

6. As a result of this ring, there is no night on Saturn in the same manner as there is on earth; there is, of course, day on the side facing the sun, but also on the opposite side, because the ring is illuminated on its inner side by the sun and the powerful light of this ring; in addition, the various orbiting moons often contribute to the illumination of the opposite side.

7. In addition to this actual "night-light" or, to better understand this circumstance, "night-day," comes a third light, which is the light of the fixed stars, which, when observed from Saturn, appears to be ten times stronger because of the planet's pure and far-reaching atmosphere, and hence the stars also give off a light many times stronger than that which Venus, the evening star, does to earth in its brightest light.

8. Now transpose yourself in your spirit onto a land in the central zone of this planet and observe from there the magnificent splendor of the starry sky! Truly, even if you were to heighten your imagination to the greatest extent, you would not be able to visualize a millionth part of the splendor which exists there, because on Saturn the night is lighter than daytime is on earth. And under the be-

nevolent shadow of the ring, you never lose sight of the beautiful sun during the day. Especially if you go to the mountains and enjoy the immense view from there, the effect of the starlight below the ring is so multifarious in its blaze of colors that it is impossible for you to imagine.

9. As far as the character of the countries in the central zone is concerned, the mountains and the rivers, the vegetation, the animal kingdom and the human beings, all of this will be revealed to you. But, for today, be satisfied with what has been given and contemplate on these matters. You will find that so far you have been given a large portion on which your spirit can and should find good nourishment. Everything else will be given to you in due time in so far as it is comprehensible to you, and it will be given in the greatest, most unrestrained abundance. You must be very diligent, because you will receive an abundance of information. That is why I say be diligent! For today I say Amen!

3
The Land Herrifa
The Healing Mountain, *Girp*
The Sun Tree, The Rain Tree
The Hair Tree, The Broad Tree And The Ray Tree

1. Regarding the countries and their character, amongst each other they are, as mentioned in the beginning, diverse in form, internal formation, and even so far as the vegetation and animals, waters, metals and stones are concerned. Not one country is similar to another in any one respect. The only exception in all the countries of this planet is the human inhabitants and the air that surrounds the planet; everything else is subject to the greatest diversity.

2. Now we will take a look at a country which, on Saturn, is called Herrifa.

3. This land is larger than Asia, Europe, Africa and the Mediterranean Sea all together. It is located a little bit above the equator of Saturn in a slanted position and has the form of an elongated egg.

4. It is the land with the highest mountains, and on the whole is more mountainous than all the others. The highest mountain is called *Girp* by its inhabitants and, according to your earth measurements, is 243,150 feet high;[4] despite its height this mountain is covered with grass and fragrant herbs even at the highest peak. It has gentle slopes throughout, and therefore the inhabitants can easily walk up this mountain, just as on a plateau. The mountain is also, at the same time, the apothecary for all the populace as well as the animals of this country. As already mentioned, you will find here the most fragrant herbs; therefore the proper healing herbs are available for whatever ails you. That is also the reason why this mountain and the surrounding area, which together comprise over 100,000 square geographic miles, is the most inhabited part of this country.

5. There are only ten different species of trees in this country. But these trees are different from those on earth because they do not produce fruit only one or twice every year; instead they are in constant bloom as well as producing ripe fruit throughout the year, year after year.

6. Amongst the trees the sun tree distinguishes itself especially. On Saturn this tree is called *gliuba*. It reaches a height of more than

---

[4] As a linear measure, 1 meter = 3.163 feet. – ED.

600 feet. The trunk of this tree quite often has a circumference that one hundred human beings from earth could not encircle. The branches quite often extend about 2,966 feet from the trunk of the tree according to your calculations and measurements. In order that they do not break on account of their weight, they produce branches of support on the underside, similar to those of the bahahania tree on earth, which are perpendicular to the ground and which, when fully grown, form the most beautiful colonnade. These supporting branches extend even from the uppermost branches, so that such a tree, when fully grown, resembles a small basalt mountain on earth, though with one difference: enough space remains between the perpendicular supporting branches that you can reach the trunk of the tree from all sides without any obstacle.

7. One leaf of this sun tree, *gliuba*, is large enough to completely cover a coachman with his horse and a large wagon. The color of the leaf is as blue as the feather of a peacock, and is adorned with the most beautiful markings. It retains its freshness and richness of color even when it is dry, resembling a ripe leaf when it falls from a tree on earth, which of course applies on Saturn as well. However, there is a difference on Saturn: such a tree is never defoliated. Because as soon as one ripe leaf falls from the tree, another grows elsewhere. The inhabitants of this area gather these leaves, which are very durable and do not tear easily; that is why they are used for the manufacture of clothing, and very artfully I might add. This clothing serves the same purpose as your overcoats. But they can also be worn directly on the naked skin, because the leaves are very gentle and soft; the top of the leaf is not as smooth as some of the leaves on your trees. These leaves can be compared to velvet. In the sunlight these leaves become particularly iridescent, almost like the tail feathers of a peacock. However, the leaves have a much more

brilliant shine than feathers. But as long as the leaves of this tree are still young, they look like polished gold which is covered with a light blue color.

8. What do the blossoms of the sun tree look like? One could say without hesitation that Solomon dressed in all his royal splendor would fade at the sight of this blossom. The blossom of this tree mostly resembles the roses on earth, but with the difference that this rose is not filled with petals; instead it forms a wide calyx, similar to bramble blossoms. Every blossom of the sun tree has thirty completely light red petals; each of these petals is about the size of a large piece of wrapping paper. The edge of each petal has a golden seam and, towards the inside of the calyx, it turns into a darker red. There are two filaments, as thick as an arm and 6 feet long, in the center of the calyx. These filaments are transparent and have the appearance of icicles in the winter. Where the anther is typically found in the flowers of earth, these two filaments have in its place two peculiar flowers which shine like a burning flame; one of these flowers shines greenish in color, whereas the other is red, but it is a lighter red than the color of the flower itself. The flower or the blossom disperses an extraordinarily exquisite fragrance. These petals and filaments are also carefully gathered by the inhabitants. The petals are used as a strengthening medicine, whereas the filaments are enjoyed as a particular favorite dish.

9. That is what the blossoms look like! What kind of fruit does the *gliuba* produce? It will be a bit difficult to give you a complete understanding of this, since there is nothing on earth with which to compare it. But to have some kind of concept of this fruit, imagine a long hexagonal fire-red stem, as thick as a man's arm, which has many stems at its end. Where this fruit is connected to the branch, there is a big knot which develops into the shape of a stem 18 inches

from the stalk. On this stem hangs a knobby fruit so large that four strong men on earth would have trouble carrying it. Inside this knobby fruit is a small plain pip, green in color and as firm as a stone, approximately the size of a nut on earth. The flesh of this fruit has the taste of bread and almonds sweetened with sugar. All the knobs or gnarls that are present in this fruit are hollow and half-filled with a juice which is comparable to the most delicious mead on earth. The color of this juice is yellow like a good old wine on earth. The color of its flesh is whitish and the outer rind has a grayish appearance at times, as if it were silver-plated but lusterless.

10. Those who live under this tree are taken care of in regards to all their needs and do not require any other property or piece of land; their share of the land is such a tree, which does not wither away or decompose, but continues to grow, more so in breadth than in height. Now the question might arise: If this tree reaches such tremendous heights, how can it be scaled in order to harvest the fruit? Even that is taken care of. The trunk of the tree as well as every branch has thorny shoots to the right and left. They almost resemble a pigeon ladder, which allows the people to scale the tree to the very top as well as to the outermost tips of the branches without encountering the slightest danger. And if it should happen that someone were to fall, no harm could come to any person on Saturn, since human beings and animals on this planet can sustain themselves in the air for a time when in dire need. Therefore, for their amusement, they can jump off the tops of the trees, which the younger generation does quite often. That such a feat is possible on Saturn is easily explained by the fact that the ring, which is at a distance of a few thousand geographic miles, divides the power of the attractive force (gravity) between itself and the planet at a ratio of 1 to 3/5. If, in addition to these weight-lightening circumstances,

there is another useful organic condition, then this difference is easily nullified and a person is placed in a position where he can maintain himself in the air for a longer period of time.

11. Now we have learned about one species of tree; nine still remain. The remaining species are not as imposing or as beneficial for Saturnites. However, these species serve the animals which inhabit this country on Saturn, and they are namely those which resemble the birds on earth.

12. Particularly notable is the rain tree, which on Saturn is called *briura*; besides the animals, human beings also derive some benefit from this tree. Like the pine tree on earth it has only one trunk, which often reaches heights of 240 feet, and it has the diameter of a medium-sized church tower. This tree spreads its branches very far and almost in the same order as the pine on earth. Its leaves are like many little whitish-green tubes, from which the purest water continuously drips. That is why the Saturnites build a large storage tank with a diameter of 600 feet around this tree, and it seems that most of these trees stand in the center of a pond. These storage tanks are built, of course, for the purpose of gathering the water which comes in plentiful amounts from the tree, and it is used by the people as well as for the few domesticated animals they keep.

13. Now you will probably say: Is there no spring water on Saturn, and especially in this mountainous area, as in the mountains of earth? Yes, there are many springs, some of which produce enough water to put your river *Mur*[5] to shame. But this spring water is considered too crude for human consumption and is therefore not used by human beings. However, as far as the Saturnites are concerned the tree water is as much as purified and similar to boiled

---

[5] *Mur* is a river in Styria (Steiermark), Austria. – ED.

water; that is why they use this tree water for all their water requirements. That is why they say: "Spring water is made only for the animals in the water and in order to water the soil. But for human beings and the nobler animals, the great God created the rain tree so that it produces well-prepared water."

14. This is the second species of tree, which, although in a much less perfect state, grows on earth intermittently, particularly in tropical countries.

15. Now we will direct our attention to the white hair tree which the Saturnites call *kiup*. This tree has a straight trunk which often reaches a height of 180 feet and is almost perfectly round. It has no branches; instead it grows silvery white threads at the top, which form a large bundle because of their abundance. The hair or threads often hang from the top to the middle of the trunk and surround the trunk at a thickness of several fathoms. These trees present a beautiful picture to onlookers, regardless of whether the wind blows or if it is quiet. A forest of these trees appears to be a forest covered with snow. The people carefully gather the threads that fall from this tree, from which they produce a kind of linen out of this material. This material is very elastic, soft and durable. This is about all which is usable from this tree and benefits the Saturnites.

16. Following this tree, we will be dealing with the broad tree, which on Saturn is called *brak*. On earth there is not one tree which resembles this one. It grows out of the soil like a golden-red wall. In the beginning it looks like a line of round tree trunks, which in time grow so close and tightly together that they turn into a real wall. A wall like this often has a length of several hundred fathoms and reaches heights of 120 to 150 feet. The wall does not have branches, twigs or leaves, but the upper edge looks like a bluish-green, thick espalier. The leaves are similar to the leaves of the platane or plata-

nus tree on earth. In the center of this espalier, fairly high pointed little stems grow forth and produce the blossoms and the actual fruit. The fruit is eaten only by birds and not by human beings; it is a kind of reddish long berry. But when the blossoms fall, the people gather them and fill sacks with them, upon which they rest; the reason for this is because of its strengthening effect and fine fragrance. A forest of these trees resembles a large maze. And when the people enter such a forest to gather the blossoms, they mark their path so that they will find their way home again. When such a group of trees is illuminated by the sun, the wall has a very strong reflection, similar to a golden surface.

17. The next tree which I shall describe is the ray tree called *bruda*. This tree is completely yellow in color, has a straight trunk, and only grows branches and twigs on the left and right side and always in a straight line. The lower branches have some kind of green stars which taper off into six regular tips. Each tip has a small blue flower, similar to a bell-flower on earth. The flower is followed by reddish fruit, which is similar to a rosehip on earth.

18. Should you wish to attain a concept of this tree, it looks like a monstrance, but like a giant monstrance. Little or nothing is used from this tree by human beings, except for ornamental purposes for tree-lined avenues.

19. As far as the other trees are concerned and a few other plants, especially the peculiar species, we will discuss these in detail in the chapters that follow. And for now, I say Amen.

4
The Funnel Tree, The Pyramid Tree And The Mirror Tree

1. The funnel tree, which on Saturn is called *kibra*, is the sixth species of tree. The trunk has a diameter of 18 feet and a smooth bluish bark. The trunk has the same diameter at the top as at the bottom, and it reaches a height of 120 feet. At the top grow branches to all sides in a 45° angle; these branches are straight and up to 60 feet long. The branches have offshoots sprouting parallel right and left, just like pine trees on earth, and the farther they are from the trunk, the longer and broader they become. The offshoots are actually the branches and the leaves of this tree at the same time. At the end of the branches you will find the blossom and eventually the fruit. Therefore this kind of tree has no more fruit than it has branches.

2. This tree has a peculiarity before its blossom time. Before the blossom has sprouted, the tree itself begins to burn; it is, however, a cold fire which resembles a glow worm and rotten wood. There is one distinct difference: this pre-blossom fire is by far lighter than the aforementioned glow worm. At such times, an entire forest of these trees gives the onlooker a spectacular light show. Not all of these trees begin to bloom at the same time; therefore the pre-blossom fire comes to some trees sooner and to others later. This fire occurs seven days before the blossoming and burns continuously, but the colors also change constantly; the result of this is that the light show of the funnel tree throughout the seven days changes from one primary color to another, inclusive of all the transitions from one color to another; in other words, the change occurs through all the seven primary colors.

3. Now imagine such a blossoming tree. Because not all the branches of this tree begin to blossom on the same day, the pre-blossom fire on one and the same tree is multi-colored. With just a little imagination you can gather how spectacular such a blossoming or, actually, a pre-blossom burning forest must appear to the eye of an onlooker (especially from a higher vantage point) when an entire forest of these funnel trees begins to blossom in this manner. At times these forests cover an area of several hundred square geographic miles.

4. Following this pre-blossom burn, the peculiar blossom of this tree makes its appearance. I am certain that this blossom would not be tolerated in many countries or states, because a six-to-twelve foot wide tri-color ribbon grows to a length of twelve feet on a twelve-to-eighteen foot long, gold-yellowish stem which is thicker than a man's arm. This ribbon has three colors of equal width, namely light red, light blue and snow-white. And as many blossoms as such a tree has, that is how many ribbon flags flutter in the wind.

5. Now you have a small concept of the splendor of the blossoms of this tree. When the blossom time is over, the flags and stems fall from the tree and the most beautiful specimens are gathered by the inhabitants. Because they lose their splendor when dry, the Saturnites roll the fallen blossoms and foliage, make piles, and, as long as they are still fresh and soft, they lie on them to strengthen their limbs. Once they become dryer and firmer, they burn them; while they burn, they exude a lovely fragrant smoke and the silvery-white ash provides an excellent fertilizer for the soil. The unattractive parts of these blossoms are left under the tree, where they decompose and thereby fertilize the soil as well.

6. The most magnificent thing about this tree is the fruit that comes into being right after the blossom. The shape of this fruit

somewhat resembles the bottle gourd on earth, but with the difference that the neck of the fruit often reaches a length of 24 to 30 feet and has a diameter of 2 feet. The neck turns at the end into a perfect sphere, the head, the diameter of which is anywhere from 9 to 12 feet. The outer rind of this fruit has the appearance of pure polished gold. Again, ask your imagination what the effect of an entire forest of such trees would be in the sunlight.

7. You probably would also like to know what this fruit is used for. The answer is very simple. For the same purpose as you use gourds on earth: partly to retrieve liquids (water) from springs or wells, partly as containers to keep or preserve juices which were pressed out from a variety of plants or fruits. This fruit is gathered as carefully as possible and kept for trade.

8. You might also like to know why this tree has the shape of a funnel. The reason for the funnel shape is that this tree can absorb the sunlight and the electromagnetic fluid more effectively. In the center of the funnel, there is a marrow tube which evaporates actual mist, especially during the night. This mist is for the other vegetation as well as for human beings; if they inhale it, it is somewhat poisonous and destructive if the light of the sun does not disperse it. However, the funnel is structured in such a way that it does not allow any more of this mist to seep through than what the tree requires throughout the night as a form of fructification or pollination, and only until the fruit has reached not quite half of its maturity. Whereupon, by approaching maturity, the marrow tube in the center of the funnel begins to close and then the mist is driven regularly to the fruit to inflate the fruit. Through this activity the funnel contains such nourishing vital air that the Saturnites climb with ladders into these funnels, prepare their beds, and sleep for many nights in these funnels.

9. Behold, that is what is memorable about the funnel tree! Besides the vital air, there is nothing which is fit for human consumption. The seeds, which are similar to pumpkin seeds on earth, are consumed only by domesticated animals.

10. Now we will discuss the seventh species of tree, the pyramid tree which is called *uhurba*.

11. This tree attains the greatest height of all trees on this celestial body, and it has almost the same attributes as your silver fir, which has a white trunk. The pyramid tree often reaches such heights that on earth there is hardly a mountain which could measure up to it. The *uhurba* has only one trunk, which at the root level often has a diameter of 480 to 600 feet. The branches start right at the bottom of the trunk and grow in every direction. The lowest branches of a fully grown pyramid tree often have a length of 6,000 feet, and, towards the top of the tree, the branches become gradually shorter and shorter, in such a manner that it assumes the shape of a mighty pyramid in the form of a round cone. When compared to this tree, your proud Egyptian pyramids are no more than little snail shells. If it were possible to physically transfer a human being to Saturn, this human being would believe himself to be looking at the highest mountains.

12. This tree belongs to the species of conifers; its leaves resemble, though of course on a larger scale, the needles of a pine tree on earth; and the color is not green but blue. The usefulness of this tree regarding the purification of the air and filling the air with vital substances is so extraordinary that the healing power from the tree tops and branches of this tree even reaches earth. The balsamic fragrant conifers on earth receive their ethereal substances from the pyramid trees on Saturn.

13. These trees are also carefully planted everywhere, which is

easily done by taking a sprig from this tree and planting it somewhere in good soil. Soon it begins to grow, and in a few Saturn years is already grown into a stately tree. It can reach an age of several hundred Saturn years. When, after this time, such a tree withers, the roots begin to decay first and the tree consumes itself even to the uppermost tops. Wherever such a tree has decayed and consumed itself, the inhabitants will spread a meager soil, where in a few years this soil becomes very fertile and will be prepared for the cultivation of their juice herbs. Even here you can employ the aid of your imagination to view several of these trees, and the result will be that the size of planet earth will begin to diminish from what you assumed it to be.

14. You now know the most important details about this tree. Therefore we shall direct our attention to the eighth species, which for you earth dwellers must truly be a highly peculiar tree. You will not find anything like it on earth.

15. The eighth species of tree is the remarkable glass or mirror tree, called *ubra*. This tree has an actual square trunk, which is as translucent as green glass on earth. The trunk tapers off with height. The tree itself reaches a height of 120 to 180 feet and has no branches throughout. More than half of the trunk of the glass or mirror tree is just like a species of cacti on earth, adorned with large hanging flowers which have almost the same shape as lilies on earth, only much larger and with the difference that on Saturn every petal of the flowers is a different color and has a total of ten petals. After half a year the blossoms begin to wither and a very peculiar kind of fruit begins to grow on a crystalline knobby stem. In the beginning this fruit consists of nothing but a translucent water pouch, which gradually becomes larger and larger. When this fruit ripens, it resembles a balloon which is 6 to 9 feet in diameter.

16. When this fruit reaches its first stage of ripening, the liquid in the pouch begins to condense to such an extent that the pouch shrinks and peels itself off little by little from the condensed liquid. The condensed liquid falls to the ground from the tree with the stem. That is when the Saturnites gather this solidified juice; they cut this fruit evenly, and the result is that they obtain square plates, which they use as mirrors. This tree is of no further use other than for adornment in their gardens, because if a row of these trees is planted it forms a magnificent avenue of splendor. The reason the inhabitants of Saturn prefer to plant this tree is because, like the pyramid tree, it is easily planted, but it cannot be propagated with a sprig. Since this tree has no branches, it is instead planted with seeds, which are contained not in the fruit but in the blossom.

17. The reason for the translucency is that its entire organism consists of square tubes, through which the useful liquid rises. If the organs were round, not one ray would be able to penetrate, because the ray would be refracted too often; whereas in this square form the ray experiences a much lesser refraction and can therefore penetrate almost without any hindrance. And because all the trees on Saturn, and by preference in this particular country, have a completely smooth and polished bark, the surface of this peculiar tree shines like mirrored glass on earth; so that any passerby can see his reflection from head to toe on the trunk of this tree.

18. This explanation covers the details of the eighth species of tree. Allow your imagination to roam in this respect, and it will not be difficult for you to understand how I, the Creator, have the ability to adorn a world without cities and palaces built by human hands. This is the information for today. We will discuss the other species of trees in the chapters that follow. Amen!

5
The Hotchpotch Tree
The Fire Tree And Oil Bush
The Physical Form Of Human Beings On Saturn
Their Real Property And Domesticated Animals

1. Tree number nine in the series of trees that grow on Saturn is the *ahaharke*. The name of this tree is difficult to translate into any language spoken on earth, because on the entire earth you would not easily find anything from which you could compose a suitable name for this tree. The only solution would be to call it "the hotchpotch tree."

2. The height of this tree is approximately 96 feet; its basic trunk also has a circumference of 96 feet. From this trunk a great number of branches grow in all directions, of which the longest reaches a length of 60 feet. From the point where these branches begin to expand or spread, three trunks rise straight up in the proper order. These trunks often reach heights of 72 to 90 feet. At the end of these trunks grow branches and twigs in the proper proportions in all directions. On each of these branches and twigs rise again three trunks to a height of 60 feet, which again grow branches and twigs in great numbers which spread and grow in the proper proportions. Above this third crown of the tree, branches again rise straight in the air, with smaller branches and twigs in proportion at the very top. When this tree is fully grown it has from seven to ten such gradations or stages, and in the order that from each trunk three new trunks rise up; in its last stages such a tree represents an actual forest.

3. Now you know why this tree is called the hotchpotch tree. The reason for this name can be given without any difficulty. The

understanding, however, is a different matter, because at every stage this tree produces different kinds of fruit and, as a result, different leaves and flowers. The most remarkable or peculiar thing about this tree, and at the same time the most unbelievable, is that it produces the same kinds of fruits only every ten years. Because from one year to the next the tree changes its fruit, and no one knows what kind of fruit the tree will produce next. And with the change of fruit, the leaves and blossoms change as well. When several of these trees are in one area, not one fruit matches the fruit of the other tree. So that the Saturnites can at all times enjoy all the fruit this tree produces, they plant ten of these trees, one every year. And whoever has ten of these trees on his property has all the products this tree produces. Because each tree bears different fruit and changes its fruit every year until the tenth year and then begins again with the fruit of ten years ago.

4. However, since each tree is different from year to year, it can happen that the first tree produces an entirely new fruit in the second year, but the tree that was planted next may produce the same fruits that the first tree produced in the first year. And when the first tree produces in the third year again another kind of fruit, then the second tree produces in the third year what the first tree produced in the second year. The third tree produces the same fruit which the first tree produced in the first year and the second tree in the second year. And that is how this order progresses. Should one of the trees wither, then another ten are planted diagonally to the old tree in order that all the fruit is available year after year and not any one fruit is missing. Furthermore, the fruit of this tree is in an order whereby the largest and heaviest fruit is naturally found at the lower level of the tree, and in that progression the fruit becomes smaller and lighter as it reaches the top of the tree.

5. The nature of each fruit of this tree and how this fruit is used by the people of Saturn cannot be described in its fullest detail, because it would require hundreds of pages if done properly. But in general it can be said that this tree is, in the noblest sense, a representative of all those tree fruits that grow in your moderate climate on earth and which contain in their center one or several well-developed seeds. For example, the lowest level would produce one year only apples, the second year only pears, the third year only plums, the fourth year only peaches, the fifth year only apricots, and so on. The higher levels of the tree produce similar fruits, but at a much more highly developed or ennobled state, also having an entirely different shape and a taste that is by far more refined and better. Whereby the fruits at the highest level are actually completely of an entirely ethereal nature and, as far as the shape and the taste are concerned, bear no resemblance to the lower levels; they are as different in earthly terms as a ripe grape is to an unripe apple.

6. And that is the progression as far as the fruit of this tree is concerned. And if you work a little bit on your imagination, you will probably be able to fill in what we have only touched upon due to lack of time, and that which was not dealt with in detail. Therefore we will conclude the description of this tree by saying that these fruits are eaten by the Saturnites, and only the fruit from the higher levels, while the fruit of the lower level is used as feed for their domesticated animals. It is of course understandable that the fruit on Saturn is ten times larger than the same fruit on earth. The bark of this tree resembles the bark of your apple tree and is also somewhat rough. But the bark of this tree is not as gray as on earth, but is dark red and becomes lighter with every level as you move upwards.

7. And now we shall direct our attention to the last tree in this order, which is the most peculiar in this country.

8. On Saturn this tree is called *fehura*, which in your language means fire tree. This tree has a similarity to what you on earth call aragonite as far as its growth is concerned; this tree is almost completely mineral. The round trunk resembles a white marble column with a circumference of 36 feet and a height of 90 to 120 feet. It has the same circumference at the bottom as it has on the top. At the top, the trunk divides, just like a little coral tree, into many different branches and twigs whose ends are all little tubes. The twigs curve and twist just like the aforementioned aragonite. This tree does not have any leaves or fruit. Its purpose is strictly that of fire. Its fruit is therefore fire, which is usually produced when a part of the country is under the shadow of the ring. On Saturn, time is not determined by seasons such as summer and winter, as it is on earth, but by the time of the shadow and the time of the light. This tree replaces the lack of sunlight at the time of the shadow through its white light. The roots of this tree are actually all little tubes which have the capability of attracting the finest natural gas. This gas is driven through the tubes into the outermost twigs, where the gas comes into contact with the atmospheric air (which at the time of the shadow contains a great amount of oxygen), ignites it and burns until the light of the sun begins to shine again. The sunlight expands the atmospheric air and precipitates the oxygen, through which the fire tree gradually becomes extinguished and the tree is at rest and will not grow until the time of the shadow sets in again. On Saturn the time of the shadow extends over a period of half a year, as on planet earth the winter season, as calculated by the temperature.

9. This tree begins to grow without seeds, like mushrooms on

earth, but not where the soil is meager; instead it grows where the soil contains the naphtha, and that is where you will most often find the fire tree. The Saturnites also plant this tree by cutting off a twig from the trunk during the time of the shadow and planting it in naphtha-rich soil. The twig will continue to burn, and through this it will grow, as well as growing roots in the soil and branching out above the soil.

10. The fire of this tree is actually not a burning fire; rather, it warms, or, rather, dispenses a warming substance over a certain distance through the effect of very intensive white rays. This is the reason by which this planet is taken care of during the time of the shadow and is therefore not much colder than it is during the time of actual sunlight. Every family seeks to plant such trees in large numbers around their dwellings and properties so that they do not lack heat or light during the time of the shadow.

11. Allow your imagination to wander when it comes to the effects of this tree, and you will certainly find that besides its magnificence, the light has a much greater effect than all your gaslights[6] if you would bring them together in one place on one light tower. Indeed, if you would plant just one fire tree on an adjacent hill, not only would it light up entire cities like ten full moons, but the entire county would enjoy its light. Now imagine many thousands of these trees throughout one country, and how their light would affect this country. If your red, ill-natured firelight invigorates you in the dark night, by how much more would such a soft white light invigorate you? However, these kinds of trees are not intended for earth, even though in the Orient, namely in some areas of the Caucasus, similar cases occur where nothing else is required but to stick a bulrush or

[6] The period referred to here is the 1840s. –ED.

a very porous piece of wood into the ground and ignite it, and it will burn like a torch, without burning the wood or the reed, though there is one difference: these flames are reddish and extremely hot.

12. This completes the tree nursery for this country and we can now deal generally with the shrubs.

13. All shrubs that grow on Saturn have the peculiarity that they are not as low as on earth; instead, they represent a smaller but, as far as the species are concerned, a very different kind of tree species. In spite of this, even the smallest shrub on Saturn grows taller and is more imposing than the most impressive trees on earth. In the country which is described here, there are over 12,000 shrubs which are all different from each other. Every species of bush has its own particular fruit, which is made very little use of except by the many inhabitants of the air. However, one of these shrubs, which is the most plentiful and is cared for by the inhabitants of Saturn, should not be without interest, because it resembles your olive tree completely, again with the difference that this shrub is many times larger. The berries of this tree are so large that, when they are ripe, one berry would produce one liter of the finest oil. And if one shrub often brings twenty to thirty thousand of these berries on its branches to maturity, this might give you an idea of the rich oil harvest, if you can also imagine that one family often has several thousand of these oil shrubs or, rather, oil trees on its property.

14. Certainly, the family property is not as small as a large farm on earth; it is at times even larger than a whole country. Besides this, the human beings that inhabit Saturn are extremely beautiful as far as their physical shape is concerned; they also taller than you are. A woman's height is anywhere from 80 to 90 feet and the height of a man is anywhere from 95 to 135 feet. Many of their domesticated animals are created in these proportions.

15. If you take these proportions into consideration from now on, everything that will be said in the proper order about the fruitful vegetation shall be much more understandable. For today, I say Amen.

6
The Herbs And The Useful Plants Of Saturn
Corn
The Flowing Barrel And The Walking Bottle Gourd

1. We have already learned whatever was remarkable in regards to the shrubs. Now we shall turn our attention to the herbs and useful plants of this country.
2. This country is one of the most mountainous on Saturn, and therefore it has the greatest number of useful, wholesome and curative plants and herbs of all possible species.
3. For example, your field crops such as grains, wheat, barley, etc., do not grow on Saturn. But in place of these there is a much more cultivated species of grain which somewhat resembles your corn, but this plant grows twenty to thirty times taller than corn on earth. Its leaves are often anywhere from 12 to 21 feet long and 5 to 8-3/4 feet wide. These leaves have a completely sky-blue color. The edges are approximately 9 inches wide, trimmed with a light carmine-red, whereas the center, which is also 9 inches wide but tapering off to 1 inch towards the tip of the leaf, is greenish-golden. The stem of this plant is as thick at the bottom as a fully grown oak on earth, and has a dark lusterless polished golden color, becoming lighter as you proceed higher up the stem. The corolla, which often has branches that spread 6 to 9 feet, looks like a chandelier with the

most brilliant, beautifully cut crystals. The reason for this appearance is because everything on Saturn is on a considerably larger scale. Should you view the corn blossom through a good microscope, you will notice almost the same kind of luster, which otherwise looks just like a whitish pollen.

4. On a larger scale, the fruit of this plant resembles a corn cob, though not in taste and usage. On Saturn the fruit is the best tasting delicacy and can be compared in this respect to a pineapple. When this fruit is ripe, the individual grains can easily be removed and eaten immediately; they are not floury; instead they are juicy, like grapes on earth. The weight of one of these berries, in accordance with earth measurements, is usually between two and three pounds. One cob often holds anywhere from 300 to 500 berries and one single plant produces 20 to 30 such cobs; this will give you an idea of how rich such a harvest can be.

5. But where do the inhabitants store such a harvest? You have learned about the good containers which the funnel tree provides. That is where the berries are stored, as well as the pressed-out juice from the berries. The fruit grows four times in one year and is extremely healthy and strengthening. The juice invigorates the hearts of the inhabitants of Saturn just as much as, if not more than, the grape and its strengthening nectar on earth.

6. Once the fruit has been harvested, the plant is left in the fields until dry; then the Saturnites drive their large draft and pack animals onto these fields where they eat the leaves, and the stems remain unharmed. Then the inhabitants cut these stems with a special saw. They make big piles all over the fields and they ignite these piles; this gives the soil the necessary nutrients for the growth and future cultivation of plants.

7. If this fruit is to thrive, the soil must be moist. But since it

very seldom rains and since there is no dew in this moderate climate, as elsewhere all over the entire planet wherever the moderate climate is predominant, and since springs are not plentiful in the countryside, what can the inhabitants do? How do they irrigate such fields? These fields usually cover an area of 30 to 40 square geographic miles. Behold, I, the Creator, solved this problem with another plant which takes care of the irrigation. This plant is planted amongst these useful plants.

8. This plant is called the flowing barrel, and it has a great resemblance to the common gourd on earth, but on Saturn these gourds reach such a size and height that even a Saturnite has difficulty looking past them. The plant itself often grows several thousand fathoms far and wide on the ground and is 6 feet thick. It proceeds from its root, having hundreds of shoots in all directions. The leaves resemble those of the pumpkin plant on earth, but they are several hundred times larger and the color is not green but violet-blue in its entirety. These leaves have markings like silvery-white stars which cover the entire leaf. The stem of the leaf is 12 to 18 feet long, round, and several fathoms in diameter. The inside of the stem is hollow; in the walls of the stem there are many thousands of tubes in the direction of the leaf; they nourish the leaf with a sweet juice; part of this liquid drips from the many pores on the underside of the tips of the leaf to the ground below and moistens the soil in this manner like a steady light rain. However, the main irrigation function of the plant is accomplished by its actual fruit. As soon as this fruit is half-ripe, at night it opens its pores on the surface, and also little tubes that protrude on the surface for this particular purpose, through which a sweet clear liquid is forced and spreads like a fountain, and the soil receives in this manner a regular and sufficient irrigation every night.

9. Now the question might arise: From whence does the fruit attain this water? The answer is that this plant is a true artesian water-well driller; it drives its roots far and deep downwards until it somewhere finds a subterranean well. There it draws the water it has found upwards with the greatest diligence and then conducts it through the best water pipes, completely purified, into all directions of its outer and fast growing and extending area.

10. Do the inhabitants of Saturn use this fruit for any purpose other than the irrigation of the fields? Yes, they use this fruit when it has fully ripened. They cut it lengthwise right through the middle. The seeds and flesh of the gourd are removed; the seeds are kept and used for future planting and the flesh of the gourd is used for feeding the cows, sheep, and goats on Saturn. The skin or peel of the gourd, which is 6 feet thick, is dried, through which it attains a great firmness. Once it is completely dry, the bottom half of this fruit is commonly used as a kind of water vehicle, whereas the top half, which is porous and has many tubes, is used in the most simple manner as a kind of wagon.

11. A hole is drilled in the middle of both walls, and through these holes a strong and well prepared axle is placed, and at each end a wheel is mounted. Another hole is drilled in the front through which a tow bar is placed leading to the axle where it is secured with a nail, and a cross-beam is mounted at the front of the tow bar. In this manner the wagon is almost ready for use, except for the wheels. These wheels do not have to be fashioned by human hands; they are fashioned by the art of nature, namely from this same plant. Nothing more is required other than the perfectly rounded stem of this gourd from which the wheels are cut off with a saw; they measure usually from 18 to 24 feet in diameter, and some of these wheels have a diameter of 30 to 36 feet.

12. This completes the work on this wagon; all that remains to be done is to tie an ox to the cross-beam or, for faster transportation, a draft dog or draft elk, and this wagon is ready for use. This wagon easily accommodates four Saturnites.

13. This kind of wagon is only used as a lighter vehicle; of course the Saturnites have much heavier and larger wagons, which they artfully build out of wood, and just like yours on earth, they are diligently outfitted with pliant and sturdy metal fittings. This metal somewhat resembles your iron; however, it is purer, longer-lasting, and does not rust. It always retains a luster like gold on the surface. The color of this metal is like platinum, which is a mixture of pure gold and pure iron. On earth a chemist would hardly succeed, following chemical procedures, in preparing or producing such a mixture.

14. Now that we have learned about these two plants, we shall direct our attention to another which is very amusing and at the same time very useful.

15. No one on earth has ever heard of such a plant; and there is no plant on earth which resembles it in the slightest, for the so-called "walking leaf"[7] of South America is actually an animal rather than a plant. The plant on Saturn which we are discussing is an actual walking plant, even though it moves from place to place like an animal. The motive force lies in its root, which has the form of a misshapen human foot, though of course it does not have any developed toes or a heel or anything else resembling part of a human foot. The whole thing is 60 feet long, a root stock bent in a right angle which has a large number of roots on all sides which catch like tentacles and suction roots. These roots cling everywhere, like the

---

[7] *Phyllium*, a genus of *orthopterous* insects; part of the *Phasmida* family. –ED.

tendrils of a vine though with one difference: these roots only remain in one place as long as they can find sufficient nourishment. Once they have taken all the moisture out of the ground, they extricate themselves from the soil and stretch farther out forward and move along the ground until they find another moist place. There they diligently drill into the ground and entwine the moist layers of earth and other herbs and grasses. Through this entwining they draw the entire plant with them. Through the activity of the foot roots, a plant of this kind travels quite a number of geographic miles in the course of a year, in accordance with earthly measurements.

16. What does this plant actually look like? The plant has a stem that is 24 to 30 feet in height, and the branches and twigs begin to grow from this stem at a height of 6 feet from the ground. Some of these branches grow downward in all directions to the ground and, in this manner, they support the plant from toppling over. These branches are usually bare of leaves; only the ones that grow upwards in multifarious curvatures from the stem produce leaves, blossoms and fruit. To some extent, this plant also resembles your grape vines, though the leaves are considerably larger and of a light blue color and their undersides are covered with little red warts. The fruit resembles the species of grapes which you call *Gaisdutte*, except that the color is not blue but as yellow as an orange and partly translucent, like green grapes. There is also a difference in the size of these grape berries. Most of them produce one liter of pure juice, and one cluster usually contains 50 to 100 berries and most plants produce from ten to twenty of these clusters. The taste of this fruit is similar to that of muscatel grapes when they are completely ripe.

17. This is the most peculiar plant on Saturn! It has a great advantage because it does not require any cultivation or labor. This

plant looks after its own needs, so that amongst the inhabitants of this planet there will be no disputes about property rights regarding this very popular plant if it should happen to enter a neighbor's property (property rights are strictly enforced on Saturn). For this reason the inhabitants grow this plant mostly in the center of their property or around the rain trees. When this plant is grown around the rain trees it remains there and does not move, because the roots are sufficiently supplied with nourishment. And should these plants be forced to move for one reason or another, they will not enter a neighbor's property right away because it would be difficult, on account of the distance, to cross over from the center of such a property to a neighbor's property. As I mentioned before, these properties are often double the size of some countries on earth.

18. The juice of these berries is used by the Saturnites for the same purpose as you would do on earth. This juice is of course much stronger than the juice which I mentioned before, nor is it kept in the same containers as the other juice; rather, it is stored in a container which is a fruit bottle especially suitable for this kind of juice; it also somewhat resembles the gourd bottle. However, this fruit bottle is considerably larger than the usual wine barrels (the so-called Heidelberg wine barrels). Once such a gourd bottle is fully grown, it can easily contain one thousand of your pails. These fruit bottles are extremely solid. The walls have a thickness of 3 to 6 feet. Once these containers are completely cleaned out, which is done by a Saturn animal, it is ready for use.

19. In the following chapters, I, the Creator, will describe more of these peculiar plants and herbs. Therefore, for today I say Amen.

7

Saturn's Wealth In The Vegetable Kingdom
Saturn's Primary Color, Blue
Aromatic Healing Herbs
The Gold Shrub, Metal Plants, Blue Grass
Form-Changing Meadow Flowers, Fragrant Algae Moss
Saturn's Mountains And Plains

1. We have already learned about some of the more useful plants. We shall deal with the other plants in a more general sense, because if we were to discuss every peculiar plant on Saturn in detail it would take too much time and would not serve the intended purpose, especially if you take into consideration that we would have to visit another seventy-six of these large countries as well as several hundred smaller islands and the entire tremendously large southern and northern areas which are covered with ice; then the many even larger countries of the ring and also the seven moons. Therefore we can only touch on that which is most peculiar, and we shall have to bypass everything else and only make mention of those which bear some resemblance to the products of earth. Herrifa, the country we have been dealing with, has countless species of plants which partly resemble those which grow on earth; but on the other hand there are many species which bear no resemblance at all to any on earth, and then again there are also plants which are native only to Saturn and exist nowhere else.

2. Regarding the plants which resemble those of earth, the principal difference is that the plants of Saturn generally surpass your earth plants by hundreds of times in size and growth. Because of their size you can see their entire magnificence with the naked eye, whereas on earth you have to employ a microscope.

3. The second difference is the color; on Saturn, green is mostly replaced by a fresh and cheerful blue in all its hues, similar to America, where some plants are more blue than green and the green itself is closer to a blue than a yellow; yellow is actually the color which is the most distant from the color of life.

4. A third difference is that the blossoms of these plants are considerably larger and more plentiful and their color meld is enhanced by something that looks like a polished metallic base which shines through.

5. Comparing the fruit of the plants on Saturn with those on earth, a single kernel of corn on Saturn is the size of a hundred or at times even a thousand of those on earth. Furthermore, a plant produces ten or even a hundred times the number of kernels. This high yield is necessary because, on Saturn, half a harvest year corresponds to fifteen years on earth. This is also why a ten-year-old Saturn child would be almost three hundred earth years of age.

6. These are the essential differences of those plants which, on a smaller scale, also grow on earth. If you want to awaken your fantasy, look at such a plant on earth and imagine all the parts of the plant to be a hundred times larger in size; in addition, imagine it with the colors that such a plant has on Saturn as well as with all the other splendors this plant possesses when viewed through a microscope; when you approach it in this manner, you will have an idea of the vegetation on Saturn.

7. In the higher mountain regions of Saturn grow extraordinary healing herbs; their ethereal aromatic healing powers are so enormous and far-reaching that not only do they keep the Saturnites in the best of health, but their healing effects reach distances of more than a thousand million geographic miles through the ether so that, for example by preference, the elder bush, juniper trees and bushes

(*Juniperus communis*) and also those healing herbs with spines[8] receive a significant amount of their ethereal healing aroma from the plants on Saturn.

8. There is one species of mountain herb on Saturn which I should explain in more detail. This herb is known by the name of *hellatharinaga*, which, when translated, means "the thousand-leaf gold shrub." This shrub grows directly on bare rock and the stem reaches heights of 18 to 24 feet. On the stem, about a thousand light red leaves wind around the shrub in a spiral form; their shape is a longish oviform (ovate), and usually 5 to 6 feet long and 3 feet wide. From the edges of these leaves protrude nine-inch long tips, so that from the stem of the leaf to the end of the leaf there are one hundred of these tips; therefore a leaf has one hundred of these tips on each side for a total of two hundred. These tips have a dark blue color which becomes lighter towards the spine tip. This spine, which is the longest in the center, has at its very tip a bundle of spines which are as red as the leaf itself. The upper side of the leaf resembles red-hot iron or burning coal that is set ablaze with a bellows, and these leaves actually produce such a fiery brilliance. The underside of the leaf is adorned with 4-1/2 inch long hair which produces all the colors of the rainbow from the leaf, and, even at some distance, you will discover a glistening rainbow under every leaf. The closer you approach to these leaves, the more the splendor of this rainbow increases, because the iridescence becomes more concentrated to the eye. The stem or stalk of the plant looks like a mat-polished gold and rises, usually by 3 feet, above the sphere of leaves,

[8] Hawthorne, thistles, etc. –ED.

upon which there are several efflorescent blossoms and upon which new buds continuously regrow.

9. The blossom has nothing in common with any flower on earth. Its shape is like a red-gold sphere with a circumference of 3 feet, to which well-shaped human arms are attached and on every arm, instead of five fingers of different sizes, there are ten spine-rays in a color similar to gold. It has an appearance almost as if someone were painting an outstretched hand, and, instead of painting a finger, he painted half a sun emanating ten rays. There are five flower petals around one flower bud which grows in a straight line from the girth of this sphere bud, so that half of the sphere is located in the calyx. In the center of this half-sphere protrude two stamens, one of which has a thickness of half a male human arm, whereas the other has a diameter of only 1 inch. The thinner one is female and the other is male; the female is white in color, whereas the male is pink. Both filaments extend 3 feet past the calyx and hang downwards towards the ground, but not touching the ground.

10. At the end of the female stamen there is a backwards-bent funnel[9] upon which the male stamen with its orifice[10] bends down too. The male stamen will from time to time drop a fragrant liquid into the funnel of the female stamen.[11] This is actual pollination or copulation of this flower. The female stamen absorbs the liquid into itself and, through this, bears the extremely powerful seeds of this plant, while the male stamen receives this ethereal liquid from the flower petals and the flower petals receive the liquid from the leaves of the stem.

---

[9] Stigma. –ED.

[10] Similar to an anther. –ED.

[11] Stigma. –ED.

11. The color of the blossom is white. Of course that makes every flower petal white, even whiter than the white of your lilies. The collar disc at the end of the flower petals looks like a polished ruby which is somewhat toothed, similar to a flat hand at the end of an arm. The rays, however, are like translucent gold.

12. This flower, or rather this healing herb, blooms and grows throughout the year, and is therefore available at all times; when here and there one is picked by the inhabitants, a new one grows in its place. When this plant is in full bloom it disperses a fragrance you cannot imagine, since nothing on earth comes close to it. The fragrance of your rose is pitiful by comparison.

13. If this healing herb, when in full bloom, would grow but once on earth, its intense extraordinary fragrance could satiate a whole country, such as Styria, with the most beautiful fragrance. Were this not the case, then how could the aromatic healing power of such a plant reach the distant mountains of other planets? That this is true can be confirmed by a person who has medial powers, and if such a person interacts astrally with this planet, he or she will tell you without reserve that the effects of this healing herb on this far-off planet can be perceived as very beneficial.

14. The inhabitants of Saturn protect this plant very carefully, and they gather or pick it very rarely, because the strengthening effect of this plant is primarily in the air which surrounds these plants. Only when such a plant reaches a ripe old age and is close to withering away (when the hairs on the leaves start to turn whitish), then it happens that the stem is carefully cut off and the seeds are sown on the rocks. The seeds of this flower are very small and resemble a fragrant dust more than seeds. This dust is drawn in by the pores of the rock, and such a plant grows out of it here and there.

15. There is one more thing we must touch upon, and that is how

such a plant takes root on a bare rock. The manner in which this occurs is that the plant spreads its roots all over the rock, similar to the stone liverwort on earth. From the larger roots that spread all over, countless fine fiber-like roots drill into the pores of the rock and hold the stem of this plant firmly so that no human being is able to tear it from the rock. The question now arises: What do these roots draw out of this dry hard rock? The answer is simple: They draw a kind of stone oil. How do they extract it from the rock? This occurs through an inherent power, which is their own smelting fire, and it makes itself known in little electric sparks which are invisible to the naked eye. However, the roots have enough power to dissolve the bordering atoms of the stone into an ethereal oil, which is immediately absorbed by the roots and is led, more and more purified, into the stalk, into the leaves and flower petals, and eventually into the ethereal seeds.

16. These are the most essential points that you should know about this highly peculiar plant of this celestial body. Even here, let your fantasy roam for a moment, and if you follow the description of this flower, you will be able to imagine it and enjoy it in your spirit and become enchanted by its healing power and splendor.

17. Of course, this is not the only healing herb or plant; there are many different plants which have healing and beneficial effects and which transmit these effects ethereally to regions on other planets.

18. Especially noteworthy are the metal plants. On Saturn they are known as *kibri*. Through this species of plant the Saturnites acquire the purest and all the various kinds of metals without the need of any chemical smelting or purification process. These metal plants come forth in the different areas of the mountains in the most magnificent forms. On earth there are plants which also contain metal; but nowhere will you find an earth plant where the roots,

stalks and leaves are completely pure metal. You may be able to artificially produce something similar when you take a piece of zinc and place it in dissolved lead and, in a short time, the so-called Saturn tree will form; it is also called a lead tree. Whatever requires a great effort on earth and can only be accomplished artificially and under the most one-sided, most tedious and poorest of circumstances, the creative power of nature works in a wide, rich and great manner on Saturn, free and without the slightest help from human science. That is why the ancient sages called this planet Saturnus; Saturnus in your language means as much as a satiated star; it also means "satiation" in all fundamental languages.

19. Behold, that is how things thrive on this planet, which is a richly blessed celestial body in every respect.

20. When you compare the growth of grass on earth to Saturn, the growth on Saturn is luxuriant and larger in size. The color of the grass is blue throughout, but passing over into a violet color. The seed stalk or funicule, which often reaches a height of 12 feet above the ground, is mostly a dazzling white, and sporadically greenish. The spikes on the stalks are mostly a light green color. In accordance with the variety of species of grasses, there is an extraordinary difference in the formation of the spikes as well as in the color and shape of the leaves.

21. The meadows are an absolute treasure of the greatest variety, with the most magnificent species of flowers. An enthusiastic botanist could not count all the species of flowers on one square geographic mile of meadow in fifty years.

22. Especially peculiar on Saturn is the *briden*. This is a species of meadow flower which changes its shape ten times in one year. Because as often as Saturn's highest moon has completed its course, and the other moons have done the same several times, that is how

often this plant changes its shape; it only takes its original shape when all the moons assume their original position; this occurs approximately ten times in one Saturn year; that is why they were given a name that means "moon flowers."

23. Besides all the different species of grasses and meadow flowers, what is noteworthy on this planet are the many species of Alpine moss. On Saturn they are called *tirbi*. They literally gild a treeless mountain height so magnificently that you can scarcely look at it when it is illuminated by the sunlight. This moss grows extraordinarily close together in different varieties, approximately 30 inches high above the rocky mountain ground. It appears throughout like an iridescent golden sandy carpet accompanied by the most fragrant Alpine scents. In these mountains, the Alpine mountaineer continuously finds himself in a very fragrant air, as if one were to enter a forest of balsam trees in the Orient in Lebanon when they are in bloom; anyone harvesting these blossoms feels as if he were in the entrance halls of heaven.

24. We have already mentioned the heights of the mountains on Saturn. There is one more thing which should also be mentioned — even the highest peaks are capable of growing some kind of vegetation, which is not possible on earth because of the necessary low air level. Also, the mountains on Saturn do not have such steep inclines, but rise like regular pyramids. Also, they do not form unbroken rows or chains, but rise above the lowlands like a meadow upon which you have piled cut grass into haystacks. And these mountains become higher and higher towards the center of the country or island. If someone were to climb the highest of these mountains in the center of such a country, he would easily be able to look over all the other mountains.

25. Now and again you find rock formations in these mountains,

but they are not as ragged as on earth. They rise on the side of a mountain in a juxtaposition like sugar loaves, some of which reach heights of anywhere from 30,000 to over 100,000 feet. However, a fully grown pyramid tree will put this rock formation to shame, because the tops of the pyramid trees reach heights higher than the peaks of these mountains. Now you can compare, from the information you have received, the height of a pyramid tree with the height of the mountains, and you will realize why this tree should be called a "growing mountain" rather than a tree.

26. This land Herrifa is the most mountainous country on Saturn, but it is also the country with the most vast plains, which are traversed in all directions by the most beautiful quiet flowing rivers which empty into the great Saturnian ocean.

27. In the chapters that follow, we will discuss what kind of vegetation grows on the banks of these rivers or streams and how they are otherwise utilized. But for today, I say Amen!

8
The Ship Plant, *Chaiaba*

1. We have already discussed the plants of this country, and the various trees and shrubbery. But before we turn our attention to the rivers and lakes, it will be necessary to learn about the plants which grow everywhere on the banks of these rivers and lakes, just like the reeds, bulrushes and various water plants on earth. We shall now discuss the ship plant, which on Saturn is known as *chaiaba.*

2. On Saturn this plant belongs to the species of trailing vines and therefore to the species of pumpkin, but there is one difference: it forms jointed sections wherever the stalk grows continuously over the surface of the ground, and at this section a large number of

white roots grow into the ground and in this manner draw new liquids and nourishment from the soil. They become more vigorous and are able to spread farther, especially along the banks of rivers and lakes.

3. What does this plant look like? What kind of fruit does it bear and for what purpose? When it emerges from the soil this plant looks like a shoot, almost like a reed on earth, one which you would use as stucco work or as a building material for your houses. The stalk grows to a height of 90 to 120 feet like a green golden pole without any leaves. At the end of this pole the plant has a blue bud which opens up gradually into a peculiar kind of blossom, just as if you were to attach ten white and light red flags to the top of a round pole.

4. These flags are attached to 12-foot, whitish-yellow straight long stems from which they unroll; and when they are fully grown they are 24 to 30 feet long and hang and flutter from these stems. The blossom which grows from these long stems is structured in such a way that it could also be considered a fruit, one which does not wither, but instead remains for years firm and permanent.

5. The pole, which is actually the stalk, usually has a diameter of 30 to 90 inches at the bottom. The inside of the stalk is hollow; however, it has the firmness of metal. Once the stalk has reached half its maturity, shoots come forth at the roots which swiftly and luxuriantly begin to meander on the ground; these shoots also have a golden-green color, only slightly lighter. These round shoots grow high stems with large broad leaves at the joint of every section. This stem is greenish-blue, round and hollow and has a length of approximately 12 to 18 feet. The leaf is an obtuse ovate and has a length of 30 feet and a width of 18 feet. The leaf is red, as red as the most beautiful roses on earth. The edge is approximately 60 inches

wide and has the color of the most beautiful light rainbow. Its surface shines like polished gold and the edges shine in majestic splendor. The underside is completely dark blue and covered with 9-inch-long hair that has the appearance of the most beautiful silk and which resembles in color the most purified indigo on earth, only slightly lighter. The petiole of the leaf is greenish-gold, as if you were to cover polished gold with a thin layer of green. The petiole is smooth and has a diameter of 30 to 60 inches at the stem. Where the petiole projects from the stalk it is surrounded by a kind of pointed crown, only that there are several tips and they are rounded at the ends and are a dazzling white color. At approximately the third section, a peculiar blossom grows on a long and strong pedicel. This blossom resembles a large bell; at the outer edge, the open end, this blossom has a diameter of 24 to 30 feet, whereas at the pedicel, the closed end, it has a diameter of 6 to 9 feet.

6. This flower is perfectly rounded in all its parts, as if it were made by a lathe. It differs in only one respect from the shape of a bell in that its wide brim is trimmed upwards with 9-inch-long tips which are joined together in a comb-like manner at regular intervals. The blossom's color is a light yellow and the tips are bright red.

7. From the center of this bell calyx emerges a dazzling white column, two times higher than the bell blossom plus the tips; in other words it grows beyond the brim. This column is the male reproductive organ, the stamen, and the tips at the brim are actually the pistils of this blossom. When the stamen is fully developed, it begins to disperse little shining stars which are attracted by the tips on the brim like electric sparks. This is actually the pollination of this plant.

8. Once the pollination has been adequately accomplished, this

massive blossom withers and falls from the pedicel without changing its form, at which point it is often gathered. It has the softness of a cushion and is used for that purpose. The tips are cut off and, because of their firmness, are used as nails.

9. How does such a fruit grow? I can tell you that it is the most peculiar fruit in the world. As peculiar as it may seem to you, in the end this plant produces a real boat. But do not assume that these are like boats or ships on earth, which can run aground with crew and cargo. That is an impossibility with these ships which are grown. And you will learn why this is so, as soon as the structure of this fruit is explained to you. After the blossoms fall off, which are just like the pumpkins on earth, the fruit below begins to develop rapidly and enormously in the following manner: if you were to build a large egg out of fine sheet metal and then dent it on the top, one girdle into the other (not one pole into the other), in such a way that the indented wall does not touch the bottom wall — in other words so that there is a space between the two walls — then that is what the fruit resembles.

10. Now apply this form to our fruit, because our fruit grows in this indented oviform and when it is fully ripened it has a length of 180 to 240 feet and a width of 90 to 120 feet. The space between the indented upper wall and the lower wall is usually between 6 to 9 feet. When the fruit is completely ripe, each of these walls has a thickness of 5 to 7-1/2 feet and the firmness of metal. When the fruit is ripe it detaches itself from the stem in which the actual seeds are located in a circular fashion. In the fruit itself is nothing but a very fine kind of air. This is why this fruit is very easy to lift; even a child can do so with very little effort. The edge is surrounded by a very peculiar molding which extends approximately 12 feet over and above the fruit and has the appearance of the fins of a fish on

earth. However, it is on all sides evenly radial and elastically firm so that no one can break anything off, unless he makes a special effort.

11. The fruit is set into the water the way it is and is used as a ship that cannot easily be destroyed. The Saturnites steer this ship with the shoot or reed,[12] and they steer it like you steer your river boats. This shoot or rudder has the advantage of being very light, and because it is hollow it is not necessary to touch the bottom with the stalk; instead, the water itself becomes the counter effect, because the cubic content of water soon becomes heavier than the hollow space of this shoot. In this manner the water itself resists the thrust of such a shoot. The flags that were also mentioned in paragraph 3 are sawed off, and with them the inhabitants adorn the railings of these natural ships.

12. Another kind of aquatic locomotion on Saturn is achieved by using the aforementioned beautiful leaves of this plant and making sails out of them. All that has to be done is to saw such a leaf off together with the petiole, as well as the pointed crown with the tips, and glue the leaf with the sap of another plant between the flags which are erected to the rim of the ship. Even a hurricane on earth would not be able to do any harm to it. This ship is able to accommodate ten Saturnites and, in an emergency, up to twenty.

13. The Saturnites very artfully connect several of these boats to make a larger boat which would make one of your great ocean liners on earth a mere toy by comparison. Because on the wide streams, large lakes and immense oceans, thousands of these boats are often connected to make one larger boat. On these boats, beautiful light-

[12] As mentioned in paragraph 3 of this chapter. –ED.

weight buildings are constructed, so that in fact such a floating ship looks more like a city than a ship.

14. Now I have told you everything about this peculiar fruit! Awaken your fantasy and you will be pleasantly surprised. The only thing that I have to add is the color of this fruit, which is very unattractive. The skin looks like the scales of a pike and is also of the same color. And for today, I say, Amen!

9

Additional Information About The Ship Plant
The Streams Of Saturn And Their Even Descent
The Sphere Shell Structure Of Saturn
And All The Other Celestial Bodies
The Wealth Of Forms Of Creation
The Four Main Streams Of Saturn

1. Since we have learned about the peculiar ship plant in the last chapter, the only thing that remains to be said is how plentiful it is and how many of these (ship) fruit such a plant produces.

2. The plant grows abundantly at the river banks and shores of lakes and the vast beaches of the oceans. In one Saturn year this plant produces fruit two times and each plant produces 400 to 500 fruits each harvest. But nobody on this planet or, rather, in Herrifa has the proprietary rights to these fruits; on Saturn, as on earth, an ancient principle of law applies: *Primo occupanti jus!*[13] Whoever is in need of such a plant or requires several, harvests the fruit; and whatever he harvests is his property. No one will contest it, and this is done for the highest moral reason, because everyone on Saturn

[13] The one who first takes possession has the rights. –ED.

considers himself to be the least and the most insignificant; this will be explained more in detail when we discuss the human beings on Saturn.

3. Now we shall take a look at the waters of Saturn, in particular on Herrifa.

4. In Herrifa there are several thousand large and wide streams which, almost without exception, have their origin in the center of this country, namely from *Girp*, the highest mountain in Herrifa. In order to understand such a possibility, you must not imagine that the foot of this mountain is as small as the mountains on earth. Instead, the foot of *Girp* covers, in all directions, an area almost larger than the continent of Europe. Now you may understand how such an immense mountain contains so many sources of water.

5. Despite its height and expansion, this mountain is shaped like an almost perfect cone; this shape is changed only by the rock formations which can be found here and there and also through the trenches that were formed by the many springs. It should be a logical conclusion that the springs which originate at such a mountain must flow in all directions; when they have reached a certain level in their descent, they are enlarged by other significant springs from other mountains, upon which they take their course towards the ocean. The difference between the rivers or streams on earth when compared to those on Saturn is that the streams on Saturn all have the same water flow or falling water, because on Saturn there is nowhere a highland to be found; there are only mountains, smaller and wider valleys and vast plains which ascend evenly without exception from sea level. And since the ascent of the land beginning from the ocean to half way up the mountain is the same everywhere, namely 6,000 feet above sea level, therefore on account of this gentle ascent all streams must have the same fall of water as far as the speed of the flow is concerned.

6. Of the many rivers or streams we will discuss only four, because they are the largest and they take their course to the ocean as straight as if you were to draw a straight line on a drawing board.

7. Where these rivers or streams originate they are already larger and wider than the Danube where it empties into the ocean. And these streams increase in width as they continue on. By the time these streams reach the ocean, each of them is approximately 200 geographic miles wide. These streams or rivers are different from those on earth in that they have a river bed which is the same depth throughout. Therefore a stream on Saturn is nowhere deeper or shallower because it has the same depth at the top of the mountains as it does at the foot. Even though on its course to the ocean the stream takes on the water of several more springs, it will become wider but not deeper.

8. You might wonder how this could be possible. I shall tell you: There is no easier explanation for this than the fact that the foundation is, everywhere and throughout, an evenly continuous rock-floor above which is an earth-layer of the same thickness, which the water carries away little by little. Under these circumstances, how could there be a difference in the depths of the river bed?

9. Because these circumstances (regarding the rivers having the same depth) might still be a little incomprehensible, it is necessary to give a brief explanation in order that you may understand this better and pay more attention to it. During the destruction of the worlds at the fall of Adam,[14] which I explained to you and which you learned about, Saturn remained unharmed in its primary structure, though it was many times larger beforehand.

---

[14] See *The Household of God*, by Jakob Lorber. –ED.

10. The former size of Saturn is represented by its present ring. The surface of the outer ring was previously the surface of Saturn. In those days, Saturn was cut evenly right and left in such a way that the northern cap and the southern cap were ejected into infinite space like two large hollow shells, because the evil serpent released a very evil brood on both these hemispheres. Only the hot center part remained pure; that is why it was preserved as a permanent reminder that the great Architect of the Universe can preserve a celestial body even though it no longer exists in its primary planetary perfection.

11. Now you probably wish to know from whence the present, much smaller planet came into being within the ring. I will tell you, but open your eyes and ears as much as you possibly can and you will get an inside look, not only as far as Saturn is concerned, but regarding all celestial bodies. The present celestial body which is now within the ring existed before the southern and northern caps were ejected, just the same as if this would happen on earth and the caps would be taken away; that is what remains. In other words, even within your earth there is a smaller earth and in this earth again another smaller one, all of them connected with each other only through air, water and fire. The present planet Saturn is actually already the third sphere, because the ring already represents two spheres, for it is split completely into two parts which do not touch each other.

12. And presently, when you look at Saturn you have the opportunity to observe a celestial body, like an apple which you have cut in the middle into two pieces. The parts which come into view show you the internal mechanical structure of a celestial body. But as far as Saturn itself in its present form is concerned, you cannot view it right to its center. However, the relationship remains one and the

same. Whatever is now in view of Saturn and farther on in the interior of the planet is formed in the same relationship, as can be seen from the surface of the outermost ring right to the planet itself in its present form. If the planet would be capped again, then another smaller ring would appear below the larger ring, and within that there would be a perfectly round celestial body floating free, as with the present one in the large ring.

13. If you raise your power of reason and sentience a little, you will understand this better and better. By thinking this through, you will realize how easy it is for Me, because it is within My power, if necessary, to either reduce or to enlarge such a celestial body.

14. If you could see all the celestial bodies in the infinite space of creation, truly you would come upon forms which your spirit would not be able to comprehend, even in the fullest light. If you human beings with your limited intellectual powers have the ability to give your own toilsome creations a significant variety of applications, then I should be able to do the same in My vast realms of creation. In this respect, My great fantasy surely would not have to go to your school in order to attend an aesthetic course of instruction, as some of your scholars seem to think.

15. The plants, animals and minerals on earth prove the wealth of your Creator's imagination in small, insignificant and meager ways. On Saturn you will already discover some of this diversity, and I can tell you it is much greater on the Sun. If what you encounter on Saturn is extremely wonderful, then what will you say when I allow you to take a look at the Sun? However, now we are dealing with Saturn, and there is yet much to see.

16. Keep in mind that as far as Saturn is concerned, we began with number one. You know that I serve the best wine last, not like bad innkeepers on earth who intoxicate the imagination of their thirsty

guests with the first glass and, instead of serving the better wine last, serve a highly diluted vinegar. Therefore understand well the significance of what I say: We began with number one! When we are finished with this celestial body, there should be some indication in your imagination and you should also have a better concept to achieve an even higher level. Because with Me the higher and always the highest to infinity never has an end. And nowhere is there a third comparison level in existence. But everywhere there is only a second; that is, there is always one above the other and one is more magnificent than the other. And nowhere is there to be found something that is the most magnificent, which will not be surpassed by something which is even more magnificent; because I Myself am the unreachable highest of all. Once you have viewed much of the splendor of the celestial bodies, only then will you be granted a faint view into heaven. And this glimpse, despite its speed, will make you completely forget all the magnificence that you have been shown on the celestial bodies. Because even though My works are infinitely exalted, these celestial bodies are in the dead, fixed matter. But by how much more will they be in the spirit, where everything is light and life?

17. But for now let us return to our celestial body, the planet Saturn, and let us measure the depth of the rivers and lakes, and also of the oceans. And a measuring tape with one and the same length will suffice, with a length of 3,000 feet. The oceans on Saturn have a depth of 3,000 feet almost everywhere, and only towards land do they become gradually more shallow. The rivers have a depth of 60 feet everywhere in the center of the river bed, and naturally it becomes more shallow towards the river banks. Only at the mouth of the rivers do the river beds gradually fall until they eventually reach the same depth as the ocean floor.

18. What happens, as a consequence of the uniformity and depth of the river beds, is that every river or stream presents to the surprised eye an almost completely smooth surface wherein the neighboring areas reflect in all their splendor, just like a quiet lake on earth. And especially at night it offers an extremely beautiful view, when the nightly light reflects an almost unimpaired shine from the rivers.

19. The four main streams divide this large country from the central mountain into four parts, so that when someone is at the top of this mountain he can see the borders of this country by following the course of these main streams. Of course a person from earth would not be able to see that far into the distance. However, the Saturnites are able to do this because their eyesight is better than some of the best telescopes on earth. And on Saturn this is somewhat of a necessity, for if someone wants to oversee his property he requires good eyesight. On a clear day, the eyes of a Saturnite have the ability to overlook an entire country from a high enough mountain on earth with ease. The Saturn being's strongest sense is eyesight; it can be somewhat compared to the eyesight of an eagle on earth, which, from a significant height, can see the smallest animals with ease.

20. Any other characteristics these rivers, lakes and oceans might have will be discussed in a future chapter. For today, Amen!

10

The People Living Adjacent To The Morning Stream
The Tree Dwellings Of Saturnites
How The Willpower Of The Saturnites Controls Nature
Saturnite Communication With The Spiritual World
The Useless Missions Of Earth Spirits To Saturn

1. Of the four main streams, there is one whose flow is exactly towards morning; this is the widest and most populated stream. But do not assume that you will find cities and fortifications erected on the banks of this stream, as would be the case on earth; you will not find anything like that on the entire planet.

2. The most marvelous accommodations for the Saturnites are the sun trees. On Saturn they are called *gliuba*, and under their many branches and trunks live entire families. The inhabitants do not all live together in one place, as is the custom on earth. Part of the population lives on the plains, preferably by the banks of streams, while another part lives in the mountains. That is why both banks of the Morning Stream are often planted with sun trees, under which the Saturnites have their permanent homes. The reason I say "permanent homes" is because these kinds of trees do not wither away after a short period of time. Instead, such trees keep on growing, becoming larger and reaching such a size that under many of them may live a single family which, through inter-relationships, often reaches a population of 10,000 to 20,000 people.

3. The rivers, but especially this Morning Stream, fulfill multifarious needs for the inhabitants, in the following order of importance: First, the inhabitants of Saturn are great friends of personal cleanliness; it is not unusual for them to wash themselves seven times a day. Secondly, they like to swim because it serves as a

particularly strengthening amusement. In this instance all of them are great artists on the water; not only do they swim on the surface, like the people of earth, but they can easily walk on the water's surface as well. Therefore they are able to perform all kinds of swimming maneuvers. For example, they can lie flat on the surface of the water or they can sit on the water's surface and twirl around on it.

4. Should they want to dive under the water, they can do this as well, but it is no easier for them than it is for geese on earth. Because in relationship to the water, the Saturnites are 40% lighter than people on earth. You can gather this kind of relationship from what has been mentioned before, namely that the Saturnites can maneuver in the air with ease, especially the younger ones. But even the older people can jump from the greatest heights without causing any physical harm. However, they do not like to do so, because after such an excursion in the air they are usually seized by an unpleasant dizzy spell.

5. Thirdly, the Saturnites prefer to live close to the water because, as aforementioned, the water emits a particularly beautiful light glimmer at night. Fourthly, these streams exhale a particularly pleasant cooling air, of which the Saturnites are great friends. Fifth, they live on the banks of the rivers because of their domesticated animals, to have easier access to water, which we will discuss later. In the mountains the larger domesticated animals cannot be kept because of the lack of adequate water, since the springs of the rivers are at the lower sections of these mountains and the higher areas have to be satisfied with the rain tree and the watering fruit which is called the flowing barrel. The largest of the domesticated animals requires, in accordance with your measurements, 598,000 quarts (566,000 liters) of water in one day, which must seem an enormous

amount to people on earth. This large domesticated animal is what you on earth would consider to be a cow; this animal is so large that an elephant could walk on the back of the Saturnian cow and be no larger than a parasite. Because of the extraordinary usefulness of this animal, the Saturnites prefer to live on the banks of streams, rivers and lakes so that this beneficial creature will suffer no harm, since it has the peculiar attribute of drinking five times as much as it eats. What this animals eats will be explained later in detail, when the actual animals are discussed.

6. Even though the Saturnites do not build cities or houses, they do know how to plant the broad tree or wall tree, discussed in a previous chapter, in such an artful manner that if you were to see such a domicile you would have to agree that it is indescribable and much more beautiful than the largest, most beautiful city on earth, and the walls look as if they were built from brightly shining polished gold.

7. Sixth, the Saturnites like to live on the banks of the streams because of their love of the sea and sailing. They do this not for selfish motives or gain, but rather for pleasure and health reasons. Also, when some people among them wish to visit neighboring or distant islands and countries, they have to make use of what is really close to their hearts — sailing on ships.

8. You might ask why the Saturnites bother to build ships when they have the ability to walk on water. The answer to this question is actually a simple one. The Saturnites can very easily walk on water as long as they are without burden. But as soon as they carry something they will sink, since the relationship of the water's capacity to carry a human being is calculated to almost a hair. And when the Saturnites walk on water, it is a very slow and cautious procedure and requires considerable practice; it requires much more

practice than figure skating on earth, whereas with their ships the Saturnites can skim over the water fairly quickly, easily traveling 30 to 50 geographic miles in one earth hour. Also, they do not have a steam power plant, paddle, or paddle wheel. Their motive force lies solely in their firm will and unshakable belief; this is also why they line the brim or railings of their ships with the flags or plant tips, which they magnetize with their will. And as a consequence the ship is drawn to the direction in which the will of the mariner has placed the "will pole" as their destination.

9. Behold, such a motive power is by far better and more unfailing than your ghastly steam engines, through which at all times the natural life of human beings is in constant danger! And if I did not take care of this through guardian angels, there would be considerably more accidents through liberated steam than there already are. Nothing is more foolish than a human being who employs forces of nature of which he has not the slightest knowledge. It is not enough to know by mere experience that the released water steam possesses a great compressive force; it is also very important to know what lies behind this liberated steam and what causes this great compressive force.

10. Dead powers are not powers; however, powers that are effective are liberated or released alive. But who knows how much power the liberated spirits in the steam possess? Truly, if they were not contained by the angels, it would not take long before all these arrogant and conceited steam engine engineers would discover how absolutely ridiculous the basis for their calculations really is. Because liberated spirits from only 1.837 liters of water, when not restrained, can change whole mountain chains into dust and ashes in the wink of an eye. This should tell you how much heavenly protection is

constantly required so that these foolish human beings with their foolish undertakings do not perish through accidents of this nature.

11. The Saturnites are unaware of any such foolishness, and despite all this they are incomparably wiser than all those scholarly super intelligent steam engine inventors, engineers and conquerors of the oceans on earth. Besides these many advantages, the Saturnites have one more invaluable benefit: that several times during their lives they can personally associate with Me and also with the angels in heaven, through which they can gain more in a brief conversation in their sphere of wisdom and cognizance than through all the excessive, stupid rhetoric of the scholars.

12. In this connection I should like to mention, in passing, that spirits from your earth quite often come to visit the Saturnites, which they are permitted to do at all times, especially when they have this need. What usually occurs is that these earth spirits, who think of themselves as being very learned, are thoroughly laughed at by the Saturnites and are reproached because of their extraordinarily foolish beliefs, for they do not even know that the Lord as the Creator of heaven and of all celestial bodies is completely human. This fact is known only by very few human beings and Christians on earth, and therefore many such spirits have the most ridiculous and absurd ideas about Me; and that is why some of them paint such nonsensical portraits of Me.

13. On some pictures I am depicted as a hierarch, while others again depict Me sitting half-naked on a cloud holding a cross in one hand, usually on the right hand of the hierarch. Others again depict Me in the form of a flying dove, whereby I must always hover above two people below Me, namely above the hierarch and above the naked, cross-carrying Christ. And others transpose Me in all three of these depictions at the same time, through which I become a mathematical nonsense, because I am portrayed as three persons of

which only two are endowed with a human form and the third is given the shape of an animal; and again these three different personalities must depict only one divine person consisting of one hierarch, one naked Christ and one dove.

14. However, on the entire planet of Saturn there exists not one human being who has any other concept of Me other than that I am completely a human being, just like any other human being, albeit with one difference: I am the most complete and the most perfect human being, that is, I am a human being in which lives corporeally the fullness of the Divinity. Should you have difficulty in understanding this, turn to matter, for it will tell you where it [the matter] comes from and what it is, and it *has* already told you so. Then it will not be so difficult to understand or to grasp what it means that in Me, as the most complete human being, lives corporeally the fullness of the Divinity.

15. Behold, when the Saturnites, so to speak, set the record straight, these conceited earth spirits become very annoyed and angry and, in revenge, they want to attack the Saturnites and impose another belief upon them by force. But the Saturnites show themselves forthwith as extremely humble but at the same time exceedingly firm in their belief, so that the spirits from earth begin to feel miserable on account of their haughtiness, and can no longer endure the sphere of the Saturnites and soon remove themselves voluntarily.

16. I shall explain such scenes in more detail when we visit the snow and ice regions of Saturn, where the departed Saturnites mainly stay and where they are active. You should know that the spirits of every celestial body, especially if they are not completely purified, mostly remain in the area of the planet which they occupied with their physical bodies.

17. But for now, we will direct our attention again to the streams.

11
The Beautiful Riparian Landscapes
The North Stream
The Evening Stream
The Midday Stream

1. If you want to properly imagine such a stream, then imagine a vast, calm water surface, which expands to the ocean in a straight line immeasurably far for your eyes. In addition, imagine that this stream flows through a vast plain which is occasionally interrupted by regular formations of mountains which are already known to us. Add to this the greatest, most abundant fruitfulness of these riparian regions, and imagine that between the mountains grow entire avenues of pyramid trees, and beautiful gardens with mirror tree avenues, and a vast forest along the embankments of these streams, with funnel trees and all other kinds of luxurious trees, shrubs, plants, herbs and grasses. Imagine the extremely peculiar population of animals and all the large, extremely diverse and magnificent water fowl which often fly about in a bevy in all directions above the vast surface of such streams, all of which are subject to the will of the Saturnites. Among the families, especially those which live on these river banks, you will also quite often find heavenly visitors such as angels and, once in a while, you will even find Me.

2. If you put this all together, you will have an approximate idea of the splendor of such a riparian region. As aforementioned, special attention should be paid to the stream that flows towards the morning with its spacious banks. But do not get the idea, especially as far as the vegetation is concerned, that everything grows in a great disarray; on the contrary, everything grows in a perfectly planned order. Not only has everything been done for animalistic

needs by a so-called casual vegetation, but I have taken the greatest care in as far as the most orderly adornment is concerned. You must come to this conclusion from the description of the plants and the entire vegetation.

3. The remaining three streams are structured in the same manner as the Morning Stream; but they do not have the width, nor are they as densely populated. Nevertheless, their splendor is not proportionately less than that of the Morning Stream.

4. The stream that flows towards the north is, along its banks, what you on earth would term "romantic," because the valley through which it flows is often confined by mountains, and on the sides of these mountains you will notice countless sky-high, dazzling white rock towers which are often adorned by the healing plant known as *hellatharianga* or the "thousand-leaf gold shrub" described earlier. This region is the best area for this plant, even though it can be found elsewhere as well, but not as plentifully.

5. Imagine the same lively banks at the North Stream as at the Morning Stream, with all the same things, save for one exception: the pyramid tree does not grow well on rocky ground. Now you have a complete picture of this stream and its banks.

6. The Evening Stream, the stream that flows towards the west, is famous because of its many harmoniously singing birds. If it were possible for you to spend an evening there, you would be spoiled by a concert like this to such a degree that the music on earth, even if it were by one of your great composers, such as Händel, played by the most magnificent Philharmonic Orchestra, would be nothing more than the croaking of frogs in a puddle by comparison.

7. Behold, consequently on Saturn I am the music teacher for the birds! You may rest assured that even your best vocalists, if they could hear such a feathered singer of this planet only once, would

not dare to sing another note for the rest of their lives. (N.B. Music is a very popular form of enjoyment for the Saturnites; the only thing is that they do not have any musical instruments. Instead, they are excellent singers, and they praise and thank Me with that talent during their divine services. All of this will be described when we discuss the actual Saturnites and their relationships and circumstances.)

8. The stream that flows towards the south, the Midday Stream, is famous for the luster of its water. The surface of the water glimmers steadily, especially during the day, just like large, beautiful, well-cut diamonds on earth; the reason for this is that this water is of a particular purity, especially on the surface. Although all water on Saturn is purer than the purest springs on earth, the water of this stream is so clear that you can see any object even at the greatest depth, as if you would see it in normal daylight; that is why the surface glimmers so beautifully, especially when there is a small movement of waves. You have no idea of the splendor — a rainbow on earth would be a very simple thing when compared to this spectacle.

9. However, as far as the population on the banks of this stream is concerned, it is without doubt the poorest. The reason for this is that the vegetation does not grow as well because of the hard water. Even though the water is extremely clean, it is harder than the water of all the other streams. Since this applies to earth, it applies on Saturn as well; the cleaner and colder a spring brings water to the surface, the harder and less fruitful is the water. This fact should not lead you to the conclusion that the banks of this stream are desolate; rather they are considerably more luxuriant than the most fruitful on earth, though not on the same level as the others on this planet, especially those of the Morning Stream.

10. Now we have viewed the four main streams from the central mountain. But these are not the only regions which are inhabited or populated. The mountains as well as all the banks of the remaining rivers are not less inhabited. These other rivers flow in part through various turns and meander to the ocean, but they mostly enter into the aforementioned main streams and other tributaries.

11. What remains now is a description of the many large inland lakes, their usefulness, their splendor, and the habitableness of their vastly expanded level shores. This will be explained at the next opportunity. For today, Amen!

12

The Inland Lakes On Saturn
The Magnificent Countryside
And The Large Population At The Shores
Links Between Rivers, Streams And Lakes
Stone Cone Groups As Places Of Entertainment
Swan Rides

1. The inland lakes mentioned in the previous chapter are quite different than the inland lakes on earth, which are gatherings of standing water at irregular depths. On Saturn the spring water also gathers in the depressions of the plains which comes from all sides off the mountains. These lakes have a much shallower bed than the rivers. Rarely is a lake bed any deeper than 36 feet, which is very shallow by Saturnian standards. Any Saturnite can easily wade through such a lake because of its shallowness. The water will hardly reach much above the knee of a Saturnite, and at times only that high. Nevertheless, it is deep enough to carry their ships.

2. The question arises: What purpose do these lakes serve?

Actually they serve the same purpose as the man-made canals on earth. These inland lakes often have up to a hundred outflowing currents or even fairly broad flow-offs that lead to various other rivers, thereby becoming tributaries. These inland lakes make it possible for the Saturnites to reach the four main streams on these waterways, which can be accomplished conveniently in the following manner. For example, if there is such a land lake between the Morning Stream (east) and the Midday Stream (south), it will have at least one if not more of these tributaries that lead to the Morning Stream as well as to the Midday Stream. You will also find a land lake between the Midday Stream and the Evening Stream (west) and also between the Evening Stream and the Midnight Stream (north) and between the Midnight Stream and the Morning Stream. In this manner the connection on water is not only established once but hundreds of times. The way in which the connection between the inland lakes and the main streams is guaranteed, there is also a connection between all the smaller rivers and streams, so that no river and almost no lake in this great country exists which cannot be easily reached by water.

3. These inland lakes cover large areas, the smallest of which is as large as the Caspian Sea (424,300 km^2) in Asia, the largest lake on earth. There are several which are larger; some cover an area that is equivalent to the Mediterranean Sea (2,496,000 km^2). But there are not too many large lakes of this size; the location of these lakes is usually in the regions that are close to the ocean. However, there are quite a number of these smaller inland lakes towards the center of this country, because wherever you find a plain in this large country, to some extent you will find in its midst some land lake of significance. When you stand upon some elevated point, this makes the view so magnificent, or, as you might say, "so picture perfect,"

that you on earth would hardly be able to imagine such a magnificent view.

4. Although on earth you have areas with inland lakes, their shape is not uniform, but is more random, and so are the surroundings. There could be weather-beaten rocks or a wooded mountain area or dirty plains and many other things that do not say much and which, unshapen, surround the area of the water or lakes. This does not apply to Saturn, because on Saturn the lakes have always more or less the form of an ellipse. These lakes have fairly large outlets which flow evenly in all directions into other inland lakes or rivers and streams. Now imagine such a calm surface of water, of which the smallest has a surface of 10 to 100 square geographic miles, then from 100 to 1,000 and from 1,000 to almost 30,000 square geographic miles in accordance to your square measure, and this will give you an idea of the dimension of these inland lakes. Then, in addition, take a look at the many wide outlets which spring from these inland lakes, most of them in a straight line, and you will truly appreciate the splendor of such a land lake, especially when you know that these outlets are anywhere from 2 to 40 geographic miles wide.

5. It is not the large dimensions alone which augment the magnificence of these lake regions; it is primarily the shores, which are highly populated around these inland lakes. Nowhere does the splendor of the vegetation develop in such grandiose unfoldment as upon the shores of these lakes. This is the environment for the pyramid trees, which, from your point of view, grow to unbelievable heights, and their tips are often higher than some of the high mountains on Saturn.

6. The Saturnites also plant the ship plant to beautify the shores. Behind these fields, where the ship plants grow and thrive, the wall

trees are grown and cultivated, which often significantly exceed the height mentioned in an earlier chapter. On some shores the golden wall of these trees grows to heights of 3,000 feet. The Saturnites have attained such a proficiency in the art of grafting that when a wall tree has grown a number of feet, they cut it and graft it into a cleave which they cut into another wall tree and then cover the graft with earth; through this, if the tree continues to grow, the trunk of this tree is already taller by one-half. This procedure is continued as long as possible, and it becomes a magnificent adornment for those shores, because these tree trunks grow to astonishing heights, far beyond their natural height. Behind these walls the Saturnites usually grow the column-like hotchpotch tree which you already know and which, on account of its extraordinary multifariousness, contributes a great deal to the embellishment of the gradually ascending shore regions.

7. When we discussed the Morning Stream, the inhabitants and their accommodation, it was mentioned that you will not find a city, fortification or any other dwelling anywhere on this planet except for this tree. This applies to the shores of the inland lakes as well. There are no dwellings, other than the first main tree we discussed.[15] There is, however, one thing that must be mentioned: when this tree grows in the regions of the shores of these inland lakes it considerably surpasses in size the circumference and height of the same species of tree that grows near rivers, streams and in the mountains, and therefore it serves more family members as a dwelling.

8. All the other trees, herbs, plants and grasses are also carefully kept in good order. Therefore these shores look like a perfect

[15] The sun tree, *gliuba*, chapter 3, paragraphs 6–10. – ED.

paradise in the actual sense of the word. Since the sun tree is the only type of dwelling, do not think that in these shore regions one such sun tree grows right next to the other; instead, it takes a considerable amount of time to travel from one dwelling to another. The shortest distance between these sun trees is approximately 10 to 20 geographic miles, but more likely anywhere from 50 to 100 geographic miles, since these properties often have the size of a whole country (in Europe), and such a property often has no more than one of these trees and at the most no more than five to ten. Only when all the members of a family no longer have enough space under the sun tree will another tree be planted somewhere close by.

9. When it occurs that some members of the family have to move because of overpopulation, then it can happen that they move to the mountains, even though they do not like to do that. But before someone plants his dwelling tree on a mountain, he will do everything in his power to find a place somewhere on a bank or shore. Only when it is impossible to find something in his neighborhood will he move to the mountains. Of course, he will look for areas in the mountains where there are one or several springs nearby, and where the rain tree (*briura*) and the flowing barrel, which are other means of water supply, can thrive. Also, the inhabitants of Saturn can no longer keep their large cows in the mountains, which gives them warm milk. Therefore they have to be satisfied with milk from the tame mountain goats which is not as sweet.

10. If they wish to enjoy the milk of the large cow, they have no other choice but to go to the plains and visit their relatives; and either by exchanging healing mountain herbs or through labor they will obtain this precious milk. Then they fill it in the aforementioned containers and walk or drive home. You will remember the wagons they have, which we discussed earlier in this book, the ones

that are made so easily from the flowing barrel which the inhabitants of the plains often call the "mountain ship."

11. Although according to your earthly concepts these dwelling trees are quite a distance apart from each other, they are close enough for a Saturnite, since he can see his neighbor's dwelling tree despite the great distance on account of his keen eyesight. Besides that, his long legs will carry him at least 10 geographic miles in fifteen minutes with the greatest of ease. And should his neighbors' house be at a greater distance, then he will travel by ship. The speed of these ships was mentioned previously.[16]

12. The only thing that remains to be explained is whether the water of such inland lakes is stagnant or flowing, and it has already been said that it is flowing. This water has such a strong current that it flows in every possible direction. However, the movement of the water in the lakes is somewhat less than in the rivers and streams. If you wish to imagine the water flow in these lakes, imagine that it flows from the center of the lake in a radial manner to as many canals as it feeds, through which it connects with the other inland lakes, rivers and streams. But here we dealing with another peculiarity or difference in comparison to the canals on earth. The water of these canals, which connect with other inland lakes, flows in the following manner: For example, in the canals which connect the land lakes, the current in Lake A flows through the canal on the left into Lake B in a long swirl, returning from Lake B on the right side back to Lake A. Therefore a boatman can travel on the left bank from Lake A towards the morning (east), whereas another boatman on the right bank can travel from Lake B to Lake A in a current which flows in the opposite direction. I should think

---

[16] Chapter 10, paragraph 8. – ED.

that your hydro-mechanics would have severe difficulties in producing this kind of water movement. However, as far as the outflow that runs into a river or stream is concerned, here there is no counter movement; instead it either flows from an inland lake into a river or from a river into an inland lake. This does not impair navigation in the least, since all water movement on Saturn is very calm and the fastest current travels no more than 60 feet per minute, whereas a normal current travels anywhere from 6 to 30 feet per minute. Also, this movement of water has very little to do with navigation, but rather it is for the sake of movement, so that the waters do not become stagnant and foul; and through their constant movement these waters emit a beneficial breath of life.

13. The beauty of these inland lakes is tremendously enhanced by the many white stone cones which are found mostly in the center of the lakes. The reason for this is that the water of these lakes is the calmest at the center, and it could very easily become stagnant and rot. Therefore care is taken through these stone cones on which the water can surge, thereby causing a friction and becoming invigorated again through the awakening of its inherent electricity. These cones are also quite often overgrown with the healing herb known to us as *hellathariänga*, the thousand-leaf gold shrub. Through its extraordinary fragrant ethereal exhalation of life, this herb makes the surface of the lake infinitely invigorating; that is why the Saturnites quite often travel by ship to these cones.

14. Sometimes there are groups of thousands of these cones in the lakes, which are particularly magnificent in appearance. If you could see them, you would believe that you are looking at one of the largest cities on the water; even Venice would be merely a toy by comparison, because most of these cones have a circumference of 2 to 3 geographic miles and a height of 12,000 to 24,000 feet. There

is enough room on a truncated cone to build a city. Now imagine a group of thousands of these cones, and that might give you an idea as to the size of such a cone city in the lake.

15. The inhabitants of Saturn work diligently with their chisels to make one or another of these cones habitable by carving steps into these cones right to the top. Here they enjoy being for days on the cones which they have prepared in this manner. Very large cones are carved out, so that they eventually have several habitable floors. Access to the upper floors is achieved through carved-out spiral stairs on the exterior of the cone, by which they can reach the upper floors. But only those cones are selected which have no plant growth whatsoever. The cones that have plant growth are considered to be a sanctuary, and the Saturnites are of the opinion that it is a sin to put a chisel on such a cone. Occasionally they are informed by the angel spirits that it is not a sin at all, but instead it is imprudent to destroy such a noble plant with a chisel. Therefore the Saturnites spare those cones which sprout vegetation out of prudent humbleness. The tops and edges of the cones that have been prepared for habitation are tastefully decorated with all kind of leaves and the flags of the *chaiaba* plant. And that is why such a group of stone cones in the center of the quiet water surface is indeed a beautiful sight, even to the Saturnites. If the people of earth could see this spectacle they would be speechless for quite a while.

16. The beauty of such a waterside, or, rather, water cone city is enhanced by the many ships that are present and therefore there are quite a few family gatherings. Also the large variety of water fowl, which are similar to swans in a great variety of colors, enliven the water surface between these cones and they also enliven large areas of the water with their many different songs. But do not imagine these birds to be the size of your swans — such a bird is as large as

a small ship. That is why the people of Saturn ride on the backs of these water birds as a kind of enjoyment; and they are carried quite speedily by these birds for a while on the surface of the water. The Saturnites train these birds as draft animals and hitch them to their ships. If you could see such a voyage, you would term it a fairy-tale spectacle, if in front of one of these ships a few hundred of these birds are swimming and pulling it. Such a voyage is only for enjoyment and it does degenerate to the exploitation of these animals, because the Saturnites are compassionate towards all creatures. Therefore, they would not use them for hard labor, because the power of their will and belief is sufficient for all things.

17. That covers everything regarding the inland lakes. We will now begin with the animal kingdom, namely with the wonderfully shaped water animals. Here, too, let your imagination roam and you will be privy to many miracles. For today, I say Amen!

13

The Shores Of The Oceans
Danger Through Storm Floods
The Moons And The Rings As The Originators Of Floods
The Lower Species Of Ocean Animals
The Giant Blue Mussel

1. To this point we have learned almost everything about the formation of the country itself, the rich vegetation and the waters and how all this can be properly used. Now we will turn from the kingdom of the elementary metallic vegetables and water sphere, which is the primary basis for the animal kingdom, to the animal kingdom itself.

2. Before we deal with the actual animals, it will be necessary to

inspect the regions of the ocean shores, which are the main housing for the greatest variety of animals. On earth, the shorelines of the oceans are the most populated regions, with a few exceptions, because of the access to the water and because it is very convenient to do business and have a transportation system on the shores of the oceans and water systems, provided these regions are not located on cliffs or reefs or sand and mud. But on the shores of the oceans on Saturn this is quite different. On Saturn no human being lives within 40 geographic miles of the shores. The reason for this is that the lowland is not inhabited up to 40 miles inward because no one is safe from sudden flooding. On earth the ocean is subject to periodical ebb and tide; much more so is this the case upon such a large planet. The tide rises in the same proportions to this planet as it does in all things on earth and everything that is on it.

3. The reason why the tide on Saturn does not rise always to the same level is because of the seven moons of Saturn which all have a considerable influence on the planet itself. This influence occurs in those times when all seven moons are on one and the same side of the planet, as a result of the different speed in their orbits; therefore the ocean water rises higher than usual on that side of the planet. Since in your case only one moon orbits one planet, it would be imprudent to attribute ebb and tide to the moon, although regardless of that the earth's moon does exert an insignificant influence. But this influence of the moon on the earth does not even account for 1 inch when the natural rise in the level of the ocean water accounts for 6 feet. However, on a planet like Saturn the natural rise of the ocean water is a determining factor; add to this the fact that one moon raises the level by 1 inch and seven moons by 7 inches. You must apply these inches in proportion, as everything else on Saturn is proportionate to earth. You will quickly come to the

conclusion that 7 inches, after you deduct all the other usual effective causes, are responsible for 420 feet. Add to this the natural rise in Saturn's ocean water of 360 feet during tide, and that will tell you how high the water of the ocean rises at times in the regions of the shore.

4. If the ring above the ocean did not have such a beneficial effect upon the waters of the ocean, the lowland and some of the plains would be endangered for thousands of miles. But a peculiar phenomenon occurs during the time of tide through the attractive force of the ring — the ocean water very seldom invades farther inward than 40 geographic miles, because during the tide the ocean forms actual mountains of water under the rings. So the water draws together, forming these mountains, as if it did not want to intrude too far inland.

5. These mountains of water greatly resemble the waterspouts on earth, but with the difference that on account of the attractive force of the ring they quite often reach ghastly heights of 100 geographic miles. Therefore, during high tide, navigation is as good as impossible. Because when a ship is seized by such a waxing mountain of water, it is lifted with such indescribable ferocity and speed that as soon as it has reached the highest peak it is thrown downward with such a thrust that one cannot speak of a happy and unscathed return. It happens, but very rarely, that in some areas the accumulation of water is so high that it almost reaches the ring.

6. Nevertheless, even the most insignificant accumulation of the ocean is very dangerous for ship and crew. Because whenever these accumulations occur, the ocean water whirls at an indescribable speed. If any ship comes within the range of such dancing mountains of water, the ship is pulled upward to the water crest in the beginning, when the whirl is still slow. The whirling increases as the water rises higher and higher; this is how it happens that a ship that

has been swept along is thrust to such tremendous depths or else destroyed by the power of the whirling water. The diameter of such a water mountain, even of medium size, is anywhere from 20 to 50 geographic miles at its base, 10 to 20 geographic miles at the middle, and 1 to 2 geographic miles at the top. The gyration of the water in the middle of such a mountain has such a speed that one gyration takes from four to five minutes, whereas at the top it takes anywhere from one to one and a half minutes. Now you can imagine the power of the thrust that this mountain has! If a ship is on the surface of the ocean and, just below the ship, the peak of such a water mountain begins to form, the ship is lifted to ghastly heights. However, if the ship comes into the whirling current of this water mountain, it is then raised to the level of a particular water wave and is thrust far away downwards.

7. This explanation has been absolutely necessary, before beginning with our description of the animal kingdom, because we have learned that on account of these circumstances the shores of the Saturnian oceans are not habitable. What will be shown, however, in this great act of nature, is the prime generation (abiogenesis = spontaneous generation) of the animal kingdom, because through this occurs a phenomenal mating act through which the atomic ether animalcule are admitted into the water, where they multiply from class to class, until they reach a level which you on earth call the kingdom of the amphibians. On this celestial body, this animal class forms an orderly transition from water animal to land animal. Therefore the shore-land is the first level, on account of the proper sequence of stages, regarding the continued development on which the water animals can cross over from the water to land. If we want to examine the animal kingdom on Saturn, we must therefore begin in the order wherein it actually has its origin.

8. Therefore the water of the oceans is the first habitat of the

animals. Which kind of animals do we discover first on this celestial body, namely in the ocean waters? Even there, the order is the same as on earth.

9. The first class of animal consists of countless amounts of extraordinarily small white worms, which are so tiny that millions have adequate room in one normal drop of water. The second class is a larger class of worm which is already equipped with two arms. Even these are not visible to the keen eye of the Saturnites. These animalcule of the second level consume many thousands of the first species every second, and through this pattern their life into its own. The third class is a kind of elongated gray worm about the size of nematodes, namely the *anguillula aceti*. This animal class is very voracious and nourishes itself from the two lower classes and thus incorporates their life into its own. The fourth class are species of worms which have two heads and already have a length of 2.14 to 2.25 mm; towards the girth they are a little thicker, so that their shape is similar to a crescent roll. This animal only consumes its predecessors. The next class differs already from the previous ones by gender. In the preceding classes no difference in gender existed. This animal, however, on account of its two heads is structured in such a way that it unifies within itself the male and female principle, indicated by its two heads. The next or fifth class already consists of a kind of four-armed reddish coleopteron. This animal has a visible length of 4.28 to 4.50 mm and a body width of about 1.07 to 2.12 mm; it is a glutton because it eats all the preceding classes in countless amounts and thereby incorporates their life into its own. That is how, in the case of a thousand such levels of such living beings, one always enters into the other until they are accepted by the class of the crustacean.

10. The class of the crustacean is just as plentiful; what appears first is the mussel, then the snail.

11. Amongst the shell fish, the one that should be mentioned is the giant blue mussel, which quite often reaches such a size that, were you to see it on earth somewhere in an ocean, you would think it was an island, covering an area of 1 to 1-1/2 square geographic miles. This mussel constitutes the final level so far as mussels are concerned. Its death is caused by little snails. The giant mussel is their source of food, and they eat their way into it from all sides. When the mussel is consumed in this manner, the shell is either thrown onto a smaller island or onto the shores of the mainland through the ebb and tide, where the inhabitants gather these valuable shells. These mussels are secured in the earth in such a way that several rain trees are planted between two mussel shells and, in these large mussel basins, the rain tree water is gathered in the most economical manner.

12. The outer side of this giant mussel is not particularly beautiful, its color being dark green. However, the inside is much lovelier; it is similar to bright shining gold covered with a beautiful azure blue. Therefore, when you look at this mussel water-basin, it is magnificent when filled with the water from the rain trees. The people of Saturn like to bathe in this water, because it has the greatest purity, as it is saturated with an ethereal fragrance. Its scent is similar to nardus oil on earth, which is also among the most fragrant on earth.

13. You will probably ask: How do the Saturnites move such a giant mussel? This is accomplished in a very simple manner. The mussel shell is not as heavy as you might imagine, because below the ring the objects are not as heavy as on any other region of Saturn, whether in northern or southern latitude. When the inhabitants of Saturn find this mussel, they open it up by using wedges

and levers. They clean the shell carefully, close it again and glue the openings all around with a special kind of water paste. Then they wait with their ships for a little tide; this then lifts the mussel, which they have secured with a strong rope to their ship, and they begin their voyage inland on one of the rivers with a speed of which you have no concept. Under these circumstances, a Saturnite uses his mighty willpower to the fullest extent. Therefore you should not be surprised that the Saturnites transport objects from place to place which you would not dare to move because of their size and weight. This will be pointed out in a much clearer manner when other opportunities arise.

14. In the next chapters we will examine the animal kingdom more closely, but for today, I say Amen!

14
The Stick Snail
The Pyramid Snail
The Disk Snail That Produces Ointments And Garden Ornaments For The Patriarchs' Coats

1. Following the great giant blue mussel, we will now deal with the snail; first with those which are in the waters, and then with those which, in accordance with their physique, are more developed and live on land.

2. In the waters there are thousands of different kinds of snails which are successively in such order that, in a biological relationship, one comes forth from the other, although it would be better expressed to say "in respect to the procreation of life."

3. As far as the preceding species of snail is concerned, they are

of a lesser interest for your curious eye, although you could write volumes of books about these species. However, if you were to attempt to describe them in detail, you would never be able to list them all. Therefore we will only describe the species of these crustaceans which are on the highest step of development and which should be of special interest to you. We shall describe the five highest species because of their wonderful stature.

4. The first highly developed species of snail is the stick snail. It is a particularly peculiar kind of snail because the whorl of the shell resembles an elongated screw. It is similar to carving a 60-foot long stick into a screw, or, in a better description, winding a long rope around this stick in such a way that another whorl could firmly mesh from the top to the bottom into the stick. Do not imagine this stick to be too thin; at the bottom, the thickest part has a diameter of 5 feet and from there it tapers off to a point, and the whorl also becomes proportionately thinner. On earth you would call a snail like that a kind of a serpentine obelisk. The reason why we called it a "stick snail" is because that is what the Saturnites call it.

5. The outer color of this snail is absolutely beautiful. The thickest part is completely rosy red, as if you would cover finely polished silver with this color. Towards the tip, the red becomes darker, with the same metallic glow, so that the rose coloration begins with the faintest red and end with the darkest red. But the color is not the splendor of this snail — there is the trim on the whorl. The long-winding abdominal belt of this snail is adorned in the most beautiful color throughout, with larger and smaller pearls. The furrow that runs between the whorls is adorned with a golden ribbon which contains beautiful arabesque figurations. That is how the house of this snail is structured.

6. The animal that lives inside this shell is less interesting; it is

merely a polyp-like worm equipped with four proboscises. Its nourishment is made up of small snails and smaller mussels, which it grabs with its lowest tentacle, crushes, and then places the squashed food into its proboscis. With the other two proboscises, this animal feels its way around in order to find something to eat and to determine if some adversaries might close in. Should this be the case, then it withdraws into its beautiful house and closes off the exit with a whitish crust. But many times this precaution is to no avail, because its enemies are the sword lobsters, which I shall describe later on. The sword-tailed crustaceans (of the order of the *xiphosura*) have the ability to penetrate this crust; they enter this house by force and consume this defenseless animal little by little, until nothing remains. These lobsters then become victim to another larger snail, which we shall discuss later.

7. The inhabitants of Saturn gather the shells of these stick snails and adorn their gardens with them. At times they use the shells as water pipes in the following manner: wherever there are springs in the mountains, they catch the water with the muzzle of the shell and cut off the tip of the tapered part; this results in a build-up of pressure, and the water streams out with force at the smaller end. Below that they place another shell, catching the water with a large muzzle, and they continue this sometimes for many miles downhill. That such a water pipe makes a very interesting impression you can very well imagine.

8. The next species of snail is the pyramid snail. The color is uniformly a golden grass green, the abdominal girth is proportionately adorned with large snow-white elliptic shapes, and the edges are trimmed in such a way as if you were to frame an alabaster plate in a smoothly polished golden frame. This snail is very large; if you compare its size to something here on earth and measure its width,

it would be considerably higher than your Stadtschloßberg.[17] The color of the animal that lives in this house is dark gray, and it has a protruding trunk the size of an immensely large elephant's. This trunk is far-reaching and extremely strong. On both sides of the trunk are two more trunks which are weaker, and on the very ends of each of these trunks there is a sharp-sighted eye. On the underside of the snail there is a strong pair of whitish flippers which enable the snail to move fairly quickly over the ice whenever the surface of the ocean freezes. When this snail travels on the ocean its house is turned upwards, and when you view this traveling snail from a distance it looks like a swimming pyramid.

9. This snail has a very vicious nature and attacks human beings by winding its trunk around its victims, crushing them and putting them in its large mouth. But the Saturnites are well aware of this and are therefore well-equipped when they want to catch this snail. They put a noose around its big trunk and close it fast, and the snail is as good as caught. Since the snail is an air-breathing animal and draws the air through its protruding trunk, it dies quickly when it can no longer inhale air through its trunk. The Saturnites know when the snail has died by the white liquid which escapes its mouth, because the liquid is already an initial sign of its internal decomposition.

10. The people of Saturn gather this liquid because of its extraordinary fragrance, which by far surpasses your ambergris on earth. As soon as the dead snail ceases to produce this liquid, they turn it over to nature. Soon the vermin of the ocean take over and within a few

---

[17] A hill (475 m) in the city of Graz, Styria. – ED.

days they completely consume the snail, that is, only the snail itself, not its hard and firm shell; at the wide opening the thickness of the shell is quite often anywhere from 24 to 30 feet. Once the shell has been cleansed in this manner, the Saturnites remove the shell from the ocean at the time of ebb, and it is taken to its destination in the same manner as the giant mussel.

11. This snail's main source of nourishment is the kind of sword-tail crustaceans that were mentioned earlier, which exist in various sizes and in great number. None are larger than the so-called ocean crab (macrocheira), but most of them are much smaller, often no larger than a grasshopper on earth. But how does this second species of snail, this pyramid snail, make a major catch of the sword-tail crustaceans? This snail catches its prey when these crustaceans are the busiest, namely when their attention is directed towards their own consumption of the stick snail. When the pyramid snail comes across a stick-snail shell filled with these crustaceans, it winds its large trunk around it and takes it to the shore and places the snail house in such way that its wide side is exposed. As soon as the crustaceans are out of the water, they begin to move out, one after the other; this is the pyramid snail's opportunity to eat them. These crustaceans are therefore at the same time a life-gathering middle animal class through which the life of one snail passes over potentized into the life of another. Between two larger animal classes there is always a smaller animal class which is ill-disposed towards the previous larger class, but which is soon consumed as tasty nourishment by the larger class that follows.

12. The third kind of ocean snail on Saturn is the disk snail. This snail resembles to some extent the nautilus snail on earth, but with the difference that your nautilus snail is of course considerably smaller, but much thicker in proportion to its platforms on both

sides than the disk snail on Saturn to its platform. The disk of this snail often has a diameter of 600 to 720 feet, but the entire thickness from top to bottom is rarely much over 18 feet. At the time of the tide this snail is on the ocean floor, whereas at the time of the ebb it always swims on the surface.

13. When the snail is on the ocean floor, it pushes its long trunk far above the surface of the water in order to inhale. Its position can be very easily determined through this, and it is usually caught while in that position. It is of course understood that this is done during medium tide, because during high tide none of Saturn's inhabitants would dare to be on the ocean. You may question why the disk snail is not captured at the time of ebb while it swims on the surface of the ocean. This animal cannot be caught while on the surface, for it can travel there at a great speed and cannot be captured without great effort. And even if someone could catch up, no one would be able to hold onto the disk, because this otherwise gentle creature will, at the slightest touch, retract all its extremities immediately and, by means of one of its flippers that extends into the water, begin to turn at such speed that no one would dare to touch this large fast-turning disk.

14. What does this extremely peculiar snail actually look like? Indeed, I will tell you: you may immerse yourself in all kinds of wonderful fantasies, but you will not succeed in properly imagining the beauty of this snail. And this is the reason why the Saturnites seek to possess such a wondrous snail, often under severe circumstances and at great danger to themselves.

15. As far as its house is concerned, this snail forms a complete circle, the opening bordering extensively along the flat pre-whorl and taking about a third of the total circle. The opening through which the disk snail arbitrarily projects its body and its wonderful

extremities is not much wider than 3 feet, and the funnel-shaped rim of the elongated opening is so well and finely rounded that it tremendously enhances the entire splendor of the house, rather than giving it the appearance of something disorganized and incomplete.

16. What does this house look like? Look and be astonished in your imagination! As far as the exterior of this house is concerned, it has a wonderful appearance. It can be compared to a house where a most artful and competent jeweler has trimmed it in a most multifarious, well-ordered fashion with all different kinds of precious gems. There is one row of diamond-like gems all around, with a weight of one pound each. Another row which follows is a row of rubies of the same weight, another one of emeralds, and so on, with all twelve kinds of the main precious gems. And in between each row is a stone border — a free space — that has the appearance of a wide gold ribbon. This band contains in an exalted form the most wonderful drawings that faithfully depict the previous group of animal species whose life is unified in this snail.

17. The end of this snail house is finished off with a gallery of upright little golden columns about 6 feet in height; this has an appearance as if a skillful sculptor had sculptured a railing around such a disk, wheel or rondel, and, where the rods are artfully sculptured, little stick snails which are connected at the top with finely wound arches. These rods are in accordance with its kind; they are gold colored, like the gold of the stick snail itself. The winding arches have a more beautiful finish than the finest shiniest polished gold. Above each arch in an artful form you will find the pyramid snail in miniature with its very own peculiar color. The railing becomes lower at the beginning of the opening of the shell and eventually ceases completely where the animal projects its main extremities for a distance of about 30 feet.

18. That is what the upper part of this snail looks like. As mentioned before, the side walls are not any higher, wider or thicker than 18 feet, and it looks like a colonnade of columns of 12 feet in height. The columns are white throughout and do not have a base or a capital; instead, they rise from the lower protruding surface to the upper ceiling. The background behind the columns is also lightly colored and resembles a rainbow. The elongated canal or rather the elongated opening of the snail shell is completely red, as sometimes on earth the clouds at sunset glow, and it has at the same time its own phosphorous luminosity which, especially at night, is not less illuminated and it appears like a cloud illuminated by the evening sun.

19. What do the extremities of the snail look like? This snail extends a kind of round sail like a peacock with its long erectile ocelated tail coverts. When there is wind on the surface of the ocean, the sail serves the purpose of propelling the snail at exceedingly fast speeds on the surface. During a calm wind the snail fans the air with this large wheel-like sail so quickly that it moves rapidly over the surface of the ocean; this motion is accelerated by the lower extremities which are in the water.

20. When this wheel-like sail is spread out it is very beautiful. Its color is pale violet. The border all around is a radiant red and self-illuminating, like clouds in the sunset glow. The entire wheel is divided into compartments at regular intervals, each of which is adorned with an excellent drawing of a stick snail, with the tapered end pointing downward. These compartments are adorned on the backside with sword-tail crustaceans in perfect order from the smallest to the largest, and these drawings are a most beautiful gold-carmine color. Every compartment forms its own arch at the edge. In the front, this arch is adorned with a true reproduction of

the disk snail itself and on the backside on a light blue background with the pyramid snail. The outer edge of the arch on the backside is a dazzling white, and it also has its own evening-cloud red luminosity, just like in the front.

21. The long trunk which the snail uses to inhale is also completely white, but with a winding red ribbon; all along the center of this ribbon are little pale greenish-gold stars. This trunk also serves the snail as an arm to procure its nourishment. The snail's nourishment consists of a kind of ocean grass which grows on the shores. In this grass there are countless little gold worms which serve the snail as supplemental nutrients. Through this supplementary nourishment the snail acquires the life of all preceding species of animals.

22. This snail already possesses its own strong instinct and sometimes produces enough intelligence so that, in certain countries, some human beings bestow upon this snail a divine reverence — this was caused by the fact that this snail, when not provoked or pursued, will rescue an object which accidentally falls into the ocean, whether that object be animal or human. Whatever this snail finds helpless on the surface of the ocean, it seizes with its strong trunk and places upon its beautiful and spacious disk and sails therewith to the shore and deposits it with its trunk on dry land. This is also why this extremely beautiful water animal has been given so many different names by the Saturnites in the different countries. Some call it the "ocean-sweeper," because it cannot bear to have anything float on the surface of the ocean; others call it the "life-saver"; others again call it the "ocean-light," and others the "living ship" or the "miracle wheel"; and there are many more such designations for this animal.

23. Except for human beings, this animal has almost no enemies. Once it reaches a certain age it dies. But when it dies, the beautiful

house loses much of its splendor. That is why the Saturnites attempt to apprehend this animal alive and then kill it, so that the splendor of this beautiful house remains. When this animal is killed it floats on the surface of the ocean. The inhabitants take it quickly with their ships on one of the rivers towards their home. Once they have reached their destination, they carefully and skillfully remove the meat, so that the compartments of the big fan or sail are not damaged. The sail is then carefully removed from the firm body of the snail and carefully stretched out. And when it has properly dried, it is treated with extremely fragrant oils through which it becomes soft and pliable.

24. From this snail-fan they tailor a type of coat. But these coats are worn only by those Saturnites, especially in Herrifa, who are held in high esteem as patriarchs. Such a fan maintains all its colors and drawings, but it loses its own luminosity.

25. The remaining flesh of this snail is almost all grease; therefore they boil it out. This fat is then mixed with fragrant herbs, and the Saturnites prepare an extraordinarily exquisite ointment which is used only by the patriarchs.

26. But what happens to the beautiful house? It is carefully pulled on land by the Saturnites and securely placed in a horizontal manner on an earth-wall built especially for the purpose, and mostly in one of the gardens of the patriarchs. This is a very pleasing sight. Many Saturnites enjoy looking at it, and on special occasions they even walk on it. However, walking on the house of the snail is done very rarely, because most patriarchs hold such garden adornment in high esteem, since on Saturn wealth is determined only by the splendor of one's garden. In order to increase the splendor of a garden, a shell of the pyramid snail is placed on one side of the shell of the disk snail. It is not unusual for one patriarch to have up to a hundred of

these adornments in a straight line in his garden, that is of both species the same amount.

27. It should suffice to say that all you have to do is to awaken the power of your imagination and you will be astonished at the magnificence and splendor of such a garden. Because one row of those diamonds with which the surface of a snail house is adorned is more valuable than the treasury of an emperor on earth, not even counting the other precious gems and radiant gold as well as all the other splendor which is contained in these gardens of the patriarchs.

28. We will discuss the other two snails in the chapters that follow. For today, however, I say Amen!

15
The Seven Snail
Practical Utilization Of Its Shell
Weight Ratios On Saturn

1. The fourth snail cannot compete in splendor with the disk snail, although it greatly surpasses it in size. The people of Saturn usually call it the "large seven snail" — not because its shell houses seven individual snails, but because the shell of this snail consists of seven straight-upward, tower-high spires which emerge as an outgrowth from the snail's elliptical shell. The main housing for the snail is completely oval, like an egg; the tapered end is always in the water, whereas the obtuse part is always pointing upwards. The whorl is not visible on the housing, because it is only on the inside of the shell. However, as far as the whorl is concerned, you will find an outgrowth of such a spire wherever it has completed one circle. These outgrowths are arranged in such a fashion that the largest of these spires emerges at the very top of the shell in the center, and

the other spires surround the largest one in a descending order. One of these outgrowths resembles a large stick snake, but with the difference that these spires are considerably longer, and they are much thicker in diameter at the bottom at the main housing.

2. The opening of the shell is completely round and is in a good ratio with the entire immense size of this crustacean. The body fills the large shell, so that the outgrowths can be pulled in at the animal's discretion. The snail projects the outgrowths if it wishes to dive underwater, and, if it wishes to be on the surface, it constricts itself with the outgrowths into the center of the shell, through which it rises to the surface of the water. The body of this snail, which pushes through the opening onto the surface of the water, is completely white and somewhat resembles a snail on earth, except that this snail has a large, long trunk in the front between its four large feelers which it uses very effectively to catch its nourishment.

3. Its nourishment consists of all kinds of sea herbs; sometimes it eats an octopus, which it tears loose by force from the ocean floor and places in its large mouth. On the two large upper feelers it has sharp-sighted eyes which, with the aid of the feelers, it can turn in any direction. When this snail discovers any prey, it quickly rushes to the place and catches it immediately, whether it be some kind of sea-weed or an octopus. In order that it can travel, it has two strong flipper-like arms under the opening with which it catches the water and, through this motion, propels itself forward.

4. How large is this snail? It has a diameter of 3,000 feet. The outgrowth in the center is higher than the highest tower on earth.[18] The diameter at the foot is often anywhere from 120 to 180 feet and tapers like a pyramid to a point at the top. The color of the shell is

[18] This book was written in 1841-42. – ED.

tinged between green and blue. From the center outgrowth run whitish-blue stripes that give the shell a tiger-like appearance. However, there are no other adornments. As far as the outgrowths are concerned, they look like a stick snail, as mentioned before, only the opening of the snail is purple.

5. This snail is considered to be a good catch by Saturnites. When the meat has been removed from the shell, the shell is moved on the water, in the manner mentioned before, to an inland location, and the pointed part of the shell is immersed into the dry soil up to the level of the opening; then this type of vessel is used as a kind of container for seed or kernel fruit.

6. Sometimes openings are installed on all sides of this snail house and on the inside of the shell a floor is installed. In this manner this snail house is used as a magnificent dwelling by the children of Saturn. The reason for this is that such a residential home can be kept very clean because of its extraordinary interior smoothness. The foundation of the floor consists of sand — in other words, the shell floor is filled with completely dry sand right up to the opening. White flat stones are placed upon the sand, always in the most beautiful order; these stones are plentiful in Herrifa. When the floor has been laid, then the building is ready, and it looks like large dome, above which rise the well-known spires. The tips are cut off so that light can enter, but also so that the fumes and the smoke from the fires that gather can escape.

7. Since this species of snail is not too plentiful in Herrifa, such houses can usually be owned only by the patriarchs who live at the ocean shores. Because of its size and extraordinary massiveness, such a snail shell is even too heavy for the strong giant Saturnites to transport far inland. As far as the massiveness of this shell is concerned, the thickness of the walls is anywhere from 24 to 30 feet. If

you take this into consideration, it might give you an idea as to its weight.

8. If the ratio of gravity were the same on Saturn as on earth, then it would be completely impossible to transport such a shell, even if there were powers superior to those of the Saturnites. That which weighs 100 pounds on earth hardly weighs 1 pound under Saturn's ring. The Saturnites can even reduce this weight by artificial means, namely through the rarefaction of air in the interior. This method is employed especially when transporting the shell of this snail. The Saturnites use the dry branches of the pyramid tree, which are rich in resin, and they push them through the opening of the shell while they are burning. Through this combustion the air in the empty housing becomes so thin that it can be moved with ease. The people of Saturn are great masters in aerostatics. All of this will be explained in more detail in the chapters that follow.

9. This concludes the chapter of the seven snail! Here too, awaken the power of your imagination a little and you will look with great amazement upon this animal as well as the utilization of its house from the view point of the inhabitants of Saturn, and you will be even more amazed if I tell you that such a building is of an indestructible firmness. Some of these shells on Saturn are older than your civilization; the older the building gets, the firmer it becomes. That is why the oldest of these buildings is held in high regard. Even though you might be amazed by this, keep in mind that these giant animals of this planet are only miniature creatures when compared to other animal species on Saturn. But they are even larger on Jupiter and unrivaled in size to those that exist on the sun. Contemplate what has been said thus far and look forward to what is in store for you! For today I say Amen!

16
The Giant, Illuminated Ray Snail
Its Appearance During Ocean Storms

1. This, the fifth snail, is the last in the order of the snails; at the same time it is the largest and in one respect the most peculiar. This snail is called "the ray snail." It is seen by the inhabitants of Saturn only before the most severe ocean storms, which we discussed earlier. Its shape is the most magnificent that you can imagine. In order to make a comparison, there is nothing on earth that is similar.

2. But to give you an idea, imagine a large cut brilliant; because this snail is so angular, the top part is more flat, whereas it is more pointed at the lower parts. The edges, which, on the upper portion, are several thousand in number, are arranged in the most beautiful order and all of them are in triangles. They look like polished golden stripes 3 feet wide, which are everywhere enclosed by a completely equilateral triangular area. The triangular plate measures 18 feet on each side, and none is larger or smaller. Only on the very top of the snail is there a larger area which is not triangular. It has thirty-two points, and is similar to a compass dial which, at its outer ends, is also trimmed with wide gold stripes. These plates are as translucent as a cut diamond and are of the same hardness. All these facets have the ability to absorb the light of the sun and the stars and then emit it at night when it becomes darker in the various refraction of rays.

3. What is the size of this snail? While it is on the surface of the ocean it could easily accommodate all the houses, streets and squares of your capital city.[19] The shell itself is 60 feet thick and from the

[19] Graz, the capital city of Styria, in Austria. – ED.

top of the shell to the tip of the bottom, is 1,800 feet. As far as the width is concerned, the diameter usually exceeds 1 geographic mile. The opening of this shell is elliptical and has a diameter of 420 feet. The snail stretches its massive head, which is very similar to that of a walrus, through this opening. It stretches its head so far over the surface of the ocean, and mostly so straight, especially during storms, that on earth it could see over high mountains with ease.

4. Despite its tremendous size, this snail has a gentle nature and does nobody any harm. Its nourishment consists of the three different kinds of food. First, water plants which are also very large in size and in abundance. Secondly, the large sea worms. Thirdly, at times it feeds on sea birds, which are a delicacy for the snail. It only eats these birds during severe storms, because during calm water this giant animal is usually in the depths of the ocean.

5. On a dark, stormy night when the snail emerges from the depths of the ocean, it often spreads such a strong light that an area of 100 square geographic miles on the ocean is completely illuminated. Imagine that during a storm the many mountains of water almost reach the sky, and you are standing on a high mountain overlooking several thousand square geographic miles of ocean, where here and there such ray snails emerge. This will give you a small idea of the spectacle this view affords the people on this planet. It becomes especially imposing when several of these snails emerge in groups and raise their long necks above the surface of the ocean while they hunt for the storm birds that fly about. However, for you, using your terminology, such a sight would be gruesome and awfully beautiful.

6. This snail is not captured by the Saturnites, because during the calm it is never on the surface of the ocean. Secondly, the shell is too heavy to transport anywhere on land for a particular purpose. This

snail reaches a very old age and usually lives 30 Saturnian years. When it dies, the entire shell falls apart and everything else decays as well in due time. The flesh is usually devoured by a kind of fish that resembles the sharks in the oceans of earth, but has an even greater resemblance to the crocodiles.

7. This concludes the chapter about the ray snail. Here, too, awaken the power of your imagination, and with the help of the information given, you should be able to achieve a proper conception of this animal.

8. In the next chapter we will deal with the third group of crustaceans, but first we will discuss the turtles;[20] the description of these will amaze you even more than the mussels and snails. For now, I say Amen!

17

More Information About Water Animals
The Largest Whale-Like Fish — *Bisorbiobiobio*
A Transformer Of The Spiritual Astral
Of Water Animals To Animals Of The Air

1. Up until now, much information has been given about this planet and its planetary structure as well as its domain or land, plants and animals, and where there was a break in sequence, while describing the outstanding species of snails. We will now briefly and

[20] In the original manuscript, there is no mention of this species of crustacean. – ED.

in a comprehensible manner continue with the description. We will not linger too long with the individual species; instead, we will discuss only the most important matters and we will offer a general overview of all the other things.

2. Consequently, we will describe in passing the animals that live in the water in general, and then turn to the inhabitants of the air. We will not linger with them for too long either, and the same applies to the land animals, in order to sooner discuss the human beings on this celestial body. Now we will return to the water animals.

3. You have been given an understanding of the large water masses and immense oceans which are on this planet, and you also know that on earth the largest and mightiest animals live in the water. You will find the same circumstances on Saturn. Naturally, the species and classes are much different and they have little or no resemblance to those on earth. We will only mention a few, first of all those that belong to the enormously large family of fish.

4. The largest of all fish on Saturn is the *bisorhiohiohio*. This fish is approximately on the same level as the whale on earth; however, as far as its shape is concerned there is a vast difference. This fish has a completely round head, 600 feet high. Therefore this head has the appearance of a ball with a diameter of 600 feet and opens backwards in the middle. This head has no teeth or gills; in both the lower part and upper part of this immense round cavity of the mouth there are two completely flat, round, hard disks. At the rearmost part or entrance to the wide throat there is a long, pliable double tongue, which this fish uses in order to pull the nourishment which has been crushed by the two disks into the throat. And of course the huge main body is connected to this immense head. A fully grown adult fish has a main body length of approximately

18,000 feet and a height of 9,000 feet, measured from the abdominal area to the back (dorsum). This fish is up to 6,000 feet in diameter at its largest circumference. In addition, this fish has a tail of 6,000 feet in length which it uses for turning and movement. On the back of this fish there are extraordinarily strong fins which are often over 600 feet in diameter. In the abdominal area there are two swimming arms similar to those of the seals or walruses on earth.

5. If you picture the *bisorhiohiohio* in your imagination, it should become clear to you that if this fish were lying somewhere on a shore on earth, and if it were to spread its dorsal fins, it could compete with some of the higher mountains on earth. Even the Saturnites call this fish a swimming mountain, a swimming island, or a swimming land; some even call it a water planet.

6. The question arises: Do the Saturnites catch this fish? The answer is no, because every person on Saturn has an extraordinary respect for this fish. When this fish sees something approaching it on the surface of the water, it opens its large spherical head and travels with great speed towards the object and crushes it with the great power and weight of its head and devours it as soon at it enters the throat. Fortunately the habitat of this fish is mostly only the polar regions, which, on account of permanent snow and ice, are much less accessible for the Saturnites than the polar regions of earth. It is therefore a rare occasion that such a fish is seen by someone on Saturn. But if it is spotted in the northern parts, which is the habitat of this fish, this is considered a bad omen. As a result of it, the people flee as far inland as possible, because they are of the opinion that this fish has been sent by the evil spirits of the ice for their demise. At the spot where such a fish has been spotted, nobody dares to set foot for a long time. This is also why the northern parts of Saturn are rarely inhabited at all.

7. You might now ask: What then is the actual purpose of this fish? This fish is the last organ of acceptance of everything that is spiritual astral in the water animals, and it distributes this spiritual astral to all kinds of animals of the air, because not only does one future kind of air animal develop in this organ in accordance with the spiritual-substantial part, but the entire future species of air animals of this celestial body originate from it, and without this fish becoming extinct because of it. In this respect this fish resembles a small planet rather than an animal, which is also a lasting organism through which countless spiritual species pass through and develop, one distinct from the other. This applies also to your whale on earth; however, regarding the general effect, the whale is far behind the *bisorhiohiohio*, because the whale on earth only prepares feathered animal species of the polar regions, whereas the Saturn whale provides the entire planet with the feathered inhabitants of the air — in other words, within this fish the souls from the water animals are transposed into all the various kinds of souls of the feathered inhabitants of the air.

8. This fish is therefore the largest and at the same time the most noteworthy of this entire planet. Besides this, there are countless classes of fish and amphibians of all kinds imaginable, different in size, form and usefulness. Besides this giant fish, there are approximately another hundred kinds which, in regards to size, measure up to your whale on earth. In order to mention and describe them all in detail would not serve the purpose of why I disclose this planet to you. However, if you were spiritually more alert it would be very easy for you to look at even the smallest detail, not only on Saturn but also on other planets, without any explanation from Me.

9. With this chapter we have come to the conclusion regarding information about the water animals of this planet, and we shall

now turn our attention to the inhabitants of the air, which will be of considerable more interest to you than all the species of water animals of which you have learned up to now.

18
Flying Insects
The Saturn Fly
The Flying Star
The Giant Butterfly *Com*
And The Utilization Of Its Plumage

1. If you look about the earth you will find, besides the many species of birds, an even larger number of classes of species of those small winged animals which are known to you under the general name of flying insects. Such creatures also exist in large numbers on Saturn, in a great variety of species and kinds. As on earth, the fly is one of the most important insects amongst them. This is also the only animal on Saturn which resembles the fly on earth in every respect, though on the lakes and rivers you find a larger kind of fly in countless numbers. This fly is bluish-white during the day. After sunset, when it is usually very active, it illuminates like a light star, almost like the lantern fly (*genera Fulgora Laternaria*) in America and in other southern tropical countries. Of course, this larger species of Saturnian fly surpasses all those on earth in the brightness of its light, because its light is completely white and it is larger than any flying insect on earth. At night the Saturnites often enjoy the playful flight of these insects when thousands of them criss-cross the air.

2. This is an animal that is noteworthy and belongs to the class of the inhabitants of the air. Another kind of insect, which is at home on Saturn and on no other planet, is the "flying star." This

animal begins its particular vital activity only at night. Its habitat during the day is the well-known pyramid tree. At night, just after sunset, it is an uplifting spectacle for the Saturnites, when at dusk thousands of such illuminated stars fly away from a pyramid tree.

3. Why is this insect given the name "flying star"? Because it has, on each side of its elongated round body, three pointed pyramidal wings that shine very brightly; when these wings are spread out it gives this insect the shape of a six-pointed star. When this insect is completely grown, it has a diameter of 9 inches and its wings shine particularly brightly during flight. Since the flying star does not stray too far from its abode, these giant trees adopt a radiating appearance throughout the night when thousands of stars buzz around in all directions.

4. Besides this glowing insect, there are a number of other insects which glow at night in various colors. However, their light is not as strong and they are considerably smaller. That is why the Saturnites pay less attention to them. There are also several larger groups of birds; by night their plumage emits a very bright light, especially when in flight.

5. Since there is not much more to be found which is of particular importance in the insect kingdom, we will immediately turn our attention to the kingdom of the birds. And while we are on this bridge of transition we will dedicate some of our attention to the butterflies.

6. Just as the butterflies on earth carry the most beautiful colors and markings on their wings, on Saturn this applies even more so. A butterfly which is known as *com* is the largest and most magnificent of all the butterflies on this celestial body. When it spreads its wings it covers 1,719 square yards of ground. Its body is approximately 120 feet long and has a diameter of almost 6 feet. Its feet are

stronger than those of an elephant on earth. Each foot has six members and is structured in such a manner that it can be extended when necessary. Its feelers look almost like two poplar trees which protrude from its head; the feelers have branches which are in regular intervals right and left of each other, similar to the arrangement of the needles on some coniferous trees. The proboscis is longer and stronger than that of an elephant on earth. And this is what this butterfly looks like: In accordance with its body it resembles a robust creature, but that is not the case at all. This creature is extraordinarily shy and is very difficult to apprehend. And this difficulty is increased significantly by its swift flight.

7. The young girls on Saturn are the best qualified and most skillful in capturing this animal, because it is much more effortless for them to float in the air than the males. For this purpose they use a pair of artificial wings and follow the butterfly with great speed. When they catch it in the air, there is great jubilation among them; they use everything the butterfly provides for decorating their clothing, because there is hardly a planet where the female gender, especially the younger generation, attaches as much value to pretty clothing as on Saturn. In order to understand why this butterfly is of so much value it will be necessary to show its splendor. It will, however, be fairly difficult to give you a proper idea of the supernatural beauty of this butterfly. It wings are completely square, except that they have a tapered tip on both sides at the ends towards the head, which is approximately 9 feet long and closely resembles a wide sword.

8. The color of the upper part of the wing looks like brightly polished rose-colored gold. On this golden surface hang a large number of the most beautiful feathers of all possible colors. These colors sparkle in a polished metallic radiance and change with the

slightest movement. The feathers are attached to the surface of the wings in such an order that the most beautiful markings and forms emerge through this. These markings are not of a permanent nature as they are on the wings of your butterflies; but the order is such that entirely different and wonderful forms appear continuously through the different movements by which the colors are changed. The edges of the wings are adorned with feathers similar to the train of a peacock on earth, except that they are much larger and more vivaciously scintillating in their blaze of color. The underside of the wings is similar to a polished gold surface, as if covered with a fine green layer of paint. The feet of this butterfly are also attired with the most magnificent feathers as well as the entire body. The feelers, however, are the most beautiful part of this animal; its main stem is extremely light and looks completely like translucent gold. If you can imagine, it reflects all kinds of different colors with every movement, almost like a cut diamond stick to which the same kind of feathers with which the edges of the wings are adorned are attached on both sides. The proboscis is a dazzling white color and sparingly wound around with ribbons that surpass a rainbow in their splendor of colors.

9. The most wonderful thing are the eyes of this butterfly. You would not be able to look into them because of their tremendous sparkle, just as you cannot look at the sun when it rises or sets. When the butterfly is killed, the splendor of its eyes ceases. That is why the eyes are not valued as much. However, regardless of this, they are carefully removed and the liquid is drained. Through skillful preparation the women make a kind of hand bag or purse out of them which, because of its translucency and durability, is used by the elegant ladies of this planet, as on earth the so-called *Ridicüle* which is either a net-like bag or knitted bag. Whatever this butter-

fly provides, nothing is discarded except for the naked inner body; everything else is used for adornment in the most extraordinary manner.

10. Why does this kind of finery have such extraordinary value? There are three reasons: first, this beautiful creature, because of its rarity, is very hard to come by; second, the feathers are very durable — the women on Saturn consider them to be indestructible; and third, the feathers are unbelievably light and retain their color and splendor.

11. There is a group of birds whose feathers are similar to those of this butterfly. These bird feathers are offered by certain profiteers on Saturn as the "real thing," but there are also well educated experts who can differentiate the real feathers from the false, the same as jewelers on earth can differentiate an imitation from a precious stone. But beware when such a swindler falls into the hands of the women to whom he sold these false feathers — they sharpen the quill of the false feathers and scratch him with it all over so that in the future he will not dare to sell imitation merchandise to someone else; furthermore, no one will buy anything more from a merchant who bears these kinds of scars.

12. This is our famous butterfly. You have learned how this animal is captured and what it is used for; it is not really necessary to mention how the women on Saturn use it as an adornment. But in passing it can be mentioned that some women who are very vain cover their entire bodies with the wings of this butterfly, so that when you look at them they look almost like the butterfly itself. This should suffice, because it is not necessary to learn more about it, and as far as I am concerned this display of vanity is no more pleasing to Me here on Saturn than it is on earth.

13. You can well imagine that besides this butterfly, there are still

countless numbers of such animals in all colors, kinds, species and sizes, and you can deduce this easily from the multifariousness of everything that is on this planet.

19
The Different Kinds Of Bats
The Flying Cow
The Flying Ribbon
The Jewelry Trade On Saturn

1. Before we begin with the actual species of birds, we will discuss a species of flying animal which on earth belongs to the class of bats (*chiroptera*) and others like it, animals with modified limbs covered with a membranous skin that serves as wings. Do we find these kinds of animals on Saturn? Most definitely so, and considerably more so than on earth. There are no bats in the actual sense; however, there are other creatures in large amounts which have such wings. If you were to deal with each of these animals in particular, you would require more than 10,000 sheets of paper. In this instance, however, it serves no particular purpose. Therefore we will discuss only a few animals of this order, and with the others we will deal only in general.

2. A particularly peculiar type of these animals is called the "flying cow." It is exceptionally beautiful and is approximately the size of a fully grown ox on earth, except that towards the tail it is about 3 feet longer; it has four feet with beautiful dazzling white claws. It is red on the back and light green in the abdominal area. The skin has a shine to it, similar to the finest silk velvet on earth. The head has a great resemblance to a greyhound, only it is a completely different color. Beginning from the neck to the head it

is light blue, and from the back of the neck to the snout it has a red stripe. The lower part of the head gradually changes to a dark blue.

3. In the area of the front feet there are two long arms right and left, which, when stretched out, cover a diameter of 36 feet. From these arms span a strong skin in connection with the hind feet; it is understood that this only occurs when the animal intends to fly; when this animal does not fly the arms are folded; each arm has three members. The arms fit so snugly to the rest of the body that they are hardly noticeable even if you are close by. However, when the flying cow spreads its wings to fly, it has the most beautiful appearance. The skin of the arms is also a dazzling white. And each arm has at its end four well-shaped fingers with strong pointed nails for the purpose of holding onto things. The winged skin looks like the finest polished gold and is adorned with light red dots and stripes in regular intervals that run into each other. The edges of this winged skin are trimmed as if with a shining rainbow which ends everywhere into completely dazzling white strings more than 30 inches long which shine like glass threads, more than the finest silk.

4. The eyes of the flying cow are extremely sharp and lively, and at dusk they sparkle like diamonds. The snout of this animal is dark red and its mouth has a fresh red color like roses. Its numerous teeth look like pure crystal. Its tongue is bright red and proportionately very long, so that this animal can make use of it for many things, such as washing its face and cleaning its entire body; because this animal has a particularly pliable body. It can use its tongue for drinking just like a dog on earth. And when this animal rolls up its tongue lengthwise, which it does all the time when it takes to the air, it produces through this tube-like tongue a very loud whistling sound which can be heard far and wide.

5. Why do they call this animal the "flying cow"? It has a full

and complete udder with four teats between its two hind legs, and this udder is filled with an extremely fine tasting milk as soon as its young are born. That is why the Saturnites often capture it, and in many areas it is tamed as a highly beneficial domesticated animal; and this is made easier because the animal is extremely gentle. When it bears its young, there is only one male among six calves. When this male is fully grown, it differs from the female in that instead of having an udder, like an sheep earth, it has a scrotum, and on its head it has one small completely white horn that is slightly bent backwards between its two down-hanging white ears.

6. If you could make use of the power of your imagination, it would not be difficult to imagine the beauty of this animal. You will probably think and say: "Why is this animal so beautiful and what purpose does it serve?" But I say to you: If you direct your attention towards the beauty of the flowers on earth as well as their multifarious forms and shapes, could you not ask in this instance as well: "Why must the blossom be so beautiful? Would it not be more suitable for such a simple seed kernel to produce a much less striking blossom?" Behold, for questions of this kind the answers have not yet matured, for regarding the beauty of such creatures it will be impossible for you to grasp the reason, since it lies in the realm of My light or wisdom. Therefore, be content solely with the idea that, by accepting the commonly valid reason for all these kinds of forms and shapes, I, the good above all and the most wise Creator of all things, know why I created all things and creatures in this manner.

7. After we have taken a look at this creature, we will take a look at other flying animals. The Saturnites call this animal the "flying ribbon" or at times the "flying rope." How does this animal come by its name? Once we have taken a closer look at it, the explanation

will become clear. Behold, the body of this animal resembles the body of a well-shaped monkey on earth. When it walks, it does so on its hind feet just like a human being. The front paws are very long and equipped with a flying membrane along the side of the body right to the middle of the hind legs; otherwise this animal uses his front paws in the same manner as a monkey. When it stands upright it is 18 feet tall, and when it squats it is half as tall. The body of the flying ribbon has nothing exceptional about it other than that the abdominal area is light blue and it has dark red wool towards the end of the back.

8. In accordance with that, what is actually distinctive about this animal? It is its long tail, which it only spreads out when it flies. When this animal walks around, it rolls up its tail so skillfully that it comes to rest right above the coccygeal vertebra, as if someone would have a tight round roll on his back. A mature animal has a tail approximately 540 to 600 feet long, with an approximate width of 30 inches. However, this tail is so fine when it is rolled up that it hardly has a diameter of 18 inches. The entire tail has so-called "internal feeling threads" which cause the rolling up of the tail. Because the tail has no bony members, this is only an extension of the skin on the back. The tail has the colors of the lightest rainbow and, on the top as well as on the bottom, it has extremely short and fine wool like uncut silk velvet, so that this wool forms many little very light glimmering wool-warts. Now you know why this animal is called the "flying ribbon."

9. You will rarely find this creature still in possession of its tail, especially in highly inhabited areas, because the Saturnites hunt it to a great extent, and it can be easily caught during the day since it does not fly during the daytime. Once it is caught, nothing will happen to this animal other than losing its tail. The tail is cut off

very close to its back and is used by the Saturnites, especially the aristocrats, to enhance one's clothing. Again, the women are the ones who adorn themselves with it, after the tail has been treated with fragrant flower oils in order to make it pliable and, like your leather, durable and tough. Usually the tail is worn as a headband, but it is also girded around the loins. This animal is always a welcome guest for the Saturnites. And since the tail grows back again, this animal is tamed and kept as a domesticated animal.

10. The jewel merchants on Saturn breed these animals. And since the price is determined by the length of the tail, they usually stitch two or even three shorter tails together and sell them as one. When such a fraud is discovered, the women will punish such jewel merchants severely.

11. On Saturn it is often the case that women exercise jurisdiction over men, since the men on Saturn are usually much in love, just as on earth. This is the reason why men are more yielding and therefore often fooled because of their love for their women; and therefore they are also at the mercy of these fashion-conscious women. On the other hand, in comparison to those on earth these women are incomparably more modest and more domesticated. This is why the men are so inclined towards them and gladly grant them many privileges. However, when we discuss the Saturnites we will deal with this aspect in more detail. Let us now return to the animals.

20

The Host Of Flying Animals Without Plumage
The Kingdom Of Birds
The Water Hen, *Behor:* A Large Species Of Heron
Heaven's Messenger
The Flight-Tone Accord Of These Birds
Song And Music Of The Saturnites

1. As mentioned, the numerous species of flying animals on Saturn are so extensive that it would take more than 10,000 sheets of paper to record their names. However, their various forms and shapes are much more admirable than their great multitude. Almost all quadruped animals on Saturn, as well as the many species of fish, also find a modification in these flying creatures. The situation is such that if all the tame and wild animals on earth, the amphibians and water animals, would have wings just like the bat, you would have winged elephants, horses, oxen, lions, tigers, hyenas and so on throughout the entire animal kingdom. What was given above as an example for earthly understanding is a reality on Saturn, with the difference that the flying animals are considerably smaller which correspond to those which have no wings and inhabit the firm ground or waters of this planet; these animals are of course also much larger in size, stronger and mightier.

2. By now you might have an idea how lively everything must be on Saturn. And the pleasant part about all this is that these creatures are mostly gentle. Through the power of their will, the Saturnites are constant masters of the elements as well as of most animals. There are only a few animals which have the reputation of the *bisorhiohiohio*.

3. Since we have taken a look at the flying animals on Saturn

which belong to the class without plumage, and we have learned how great their numbers are and their multifariousness, the thought should emerge and should result in the following question: "If there are so many flying creatures on this planet, how will they all be able to exist? If all of these animals would take to flight, the entire air space must be occupied." However, this concern on your behalf for this large planet is completely in vain. Think about one fact only: this planet is over a thousand times larger than the earth, and, as you know, it has over 70 continents, several of which have the same surface area as the entire earth's surface, even if all the oceans and all the other waters were solid land. On earth, human beings can walk around, even though in the air, on land, in the ground and in the water there are millions of creatures of all kinds. And just as people on earth are not too much bothered by animals, the same applies to the inhabitants of Saturn. Irrespective of the fact that there are so many different and peculiar species of animals on this planet, you will see even fewer animals roaming about free in nature than on earth, where everything moves about in much smaller circles than on Saturn.

4. In order that you may have an idea of the vastness of this planet, I should like to draw your attention to what has been mentioned earlier about this celestial body regarding the distance between the dwellings of the inhabitants on Saturn. As far as the housing of the inhabitants is concerned, the same applies to all the other circumstances; everything has its completely sufficient space or room. That is why border disputes are as good as unknown on this celestial body.

5. This fact must be mentioned in advance so that you do not fall victim to disbelief when I give you an account of all the feathered inhabitants of the air and the animals which inhabit the land.

6. Now we will turn our attention to the birds! You know how multifarious this species is, even on your small planet, when you begin your account with the large ostrich and continue all the way to the small hummingbird. But how small is the earth's surface in comparison to the much larger surface of Saturn! On Saturn there are thousands of times more species of these types of animals than on earth. If you wish to determine the number of species, I can tell you: if on Saturn there were only one male and one female of each species, there would be 240 million birds. Naturally, there is not one genus of bird alone in one and the same land, but there exists also in each country other species. The species that inhabit the southern part of a country do not resemble those of the northern parts, even though they belong to one and the same species. For example, there is a water hen or coot which is very well-known on Saturn. It looks entirely different in southern water than it does in northern. All species of birds, tame and wild, are different in shape, in color and usefulness — from south to north and from east to west, in one and the same country.

7. From what has been said thus far you can gather that it is impossible to record all the names of all these animals in one lifetime. You must also concede that it would be an even greater impossibility to describe every individual bird in accordance with its species, all its functions, form and shape as well as its determination. Therefore, we shall only briefly describe the first and the largest bird on Saturn from the feathered kingdom of the animals.

8. The name of our bird is *behor* or the "air ship." You can believe that if this bird were on earth it would take up more space than the largest ocean liner, without even taking its wing span into consideration. When this bird is in flight or has its wings spread, in

accordance with your measurements, the tips of the two outermost flight feathers are a good hour's distance from each other.[21] The quills of the flight feathers have a larger diameter than the trunk of the largest oak tree on earth. And each of the feathers on the wings has a length from the quill to the outermost tip of approximately 4,800 feet. This bird has also very long and strong legs. When it is standing on the ground its feet are proportionately just as long as a heron's on earth. Why does this bird have such disproportionately long legs? Because it is an aquatic bird and therefore its habitat is the ocean shore, where it lives on fish. You will never see it on land — it either swims on the water or flies low above the water; that is why the Saturnites call it the "flying ship."

9. Is this bird beautiful? The answer is no, it is not plagued by beauty. If you imagine a heron, but on a much larger scale, you will have approximately the shape of our flying ship. Its color is ash-gray throughout and at times dark brown; its beak resembles that of a goose and its head is about the same, only proportionately larger. A fish in the waters on Saturn that has the size of a fully developed shark on earth is devoured by this bird with the same ease as if you would eat a strawberry. This will give you an idea of its size and shape.

10. Some of you might ask: Is this giant bird dangerous in any way? No, not at all. The *behor* is very fearful and will flee when approached by a human being, even a child. Its size is deceptive, because it has very little to do with that bird's actual strength; only its large, many fathom-long feathers make it look so large. If robbed

---

[21] 3.75 km. – ED.

of its feathers, it would not weigh by far as much as the weakest woman on this planet.

11. Now we have learned about one bird, the largest on this planet. This bird is also quite different at the various oceans in regards to size, color and shape.

12. The next bird in line is the most peculiar on Saturn and is known as "heaven's messenger." This bird resembles a white pigeon on earth as far as shape and color are concerned, though it is 500 times larger. The Saturnites believe that this bird flies constantly in the air, because nobody has ever seen heaven's messenger sitting or resting anywhere. The Saturnites are right in one respect. Because the heaven's messenger does not rest anywhere on land, it flies about very leisurely, high in the air at times and then low. But when it becomes hungry it flies with great speed towards the ocean shores, where it hides in the remotest corners and seeks its nourishment there, which consists of a kind of fatty white cliff moss.

13. Once the bird has eaten and regained its strength, it flies inland again to extraordinary heights. This it usually does early in the morning before sunrise, which is one reason why it is also called the "sun messenger" in some areas of Saturn.

14. While in flight this bird sings the most beautiful songs, in a much more perfect manner than the nightingale on earth. This is why this bird is also called the "merry morning singer," especially by the Saturnian women.

15. Even though this white bird is mostly sighted and heard quite frequently in parts of the country which are situated close to the ocean, the people still stand and watch it until they lose sight of it in the distance. At times the Saturnites are so taken by this bird that they would bestow upon it a divine reverence if this were allowed by the spirit-angels of the planet.

16. But this will not happen, for these birds have an inborn instinct. They avoid nothing more than the gaze of a human being. If a person on Saturn merely envisages such a bird, he may be assured that the bird will soon withdraw this person's desire to view it. That is why this bird inhabits places which are inaccessible to the view of the Saturnites.

17. Especially noteworthy is this bird's sometimes extraordinarily fast flight; it is not that easy to gain the right concept of its speed. Once this bird has reached a certain height, it easily travels 1,000 geographic miles in one hour. When this bird is in flight during the night, it glistens throughout in a bright white color. And because it flies at high speeds, it almost has the appearance of a shooting star. It prefers to fly across land at night, when it becomes a major spectacle for Saturnites. Some Saturnites are so fond of these light-apparitions that, in those regions where this bird is at home, they lie on their backs on a hill without trees so as to have an unobstructed view of such a bird in flight.

18. There is one other peculiarity: if two or three birds fly in a straight line at high speeds, they produce a pure sound by cutting through Saturn's air. When several of these birds fly in this manner in the same direction, almost each bird produces another sound. These sounds often form a chord which swells from a pianissimo to a fortissimo and fades again to a pianissimo, the same as a tone or chord on the piano.

19. This is why this bird holds many extraordinary attractions for the Saturnites. Although they are very fond of singing and especially of harmonious chords, they are not very musical. They have the simplest and most primitive musical instruments, but they have much better singing voices. The women usually sing the melody and the men sing only the chords. These singers often amuse themselves

for days with one chord, after they are fortunate enough to have found one. However, it is very difficult for them to find another good or harmonious chord after an interruption. As far as the Saturnites are concerned, any other circumstances will be presented in a clear manner at the right time. Since we cannot report anything more of great importance in regards to our heaven's messenger, sun bird and merry morning singer, we shall turn to another of our feathered inhabitants of the air.

21

The Singers Above The Rivers And Lakes
The Masters Of The Art Of The Fugue
The Northern Singer Of The Air,
The Most Effective Music

1. The species of birds which we will now discuss are called the "singers above the rivers and lakes." These birds were mentioned once before because of their delightful singing. However, we will dedicate a little more attention to them and first take a look at their shape. These birds somewhat resemble the swans on earth, only twenty to thirty times larger in size, and the neck is not as long proportionately, but instead is much thicker. The head is proportionately larger than the head of a swan.

2. These birds have a very flexible larynx which is in contact with a very agile tongue and in proportion to the rest of the body they have large, very elastic lungs with a large air capacity. They are the actual musicians on this planet — as far as music is concerned, they are entertainers. This bird never repeats anything it sings. Even if it sings for years, it will never repeat a melody it has already sung.

3. This, however, is not what is surprising as far as the musical

art of these aquatic singers is concerned, but when several of these birds gather (which is what usually happens because they like to sing in company), you will never hear one disharmonious chord. Should one of these birds begin to sing, it will not take long for a second, third, fourth and so on to join in, but it will never be one and the same melody. In spite of this, each bird will sing its own particular melody, but, guided by its very sensitive feelings, the singing of the other birds will never be in disharmonious contrast, even if there are up to thirty or more.

4. The singing of these birds would be a feast to the ears of every friend of the strictest and most accomplished fugue. Not only is there a continuous meeting of new ideas, but these ideas are modulated and change the fundamental tone so suddenly that even the greatest composer on earth would not be able to imagine it. And imagine that the purest of voices are added to all of this; the voice of the best singer on earth would be a mere shriek by comparison. This will give you an idea of what great enjoyment these birds represent to the Saturnites, who from birth are great friends of music. All I can tell you is that if it were possible to hear only three tones from the throat of such an aquatic singer, all the music on earth would be repulsive to you for the rest of your life.

5. These aquatic singers are also the reason why the Saturnites themselves do not show too much interest in the subject of music, even though they are great friends of music. They say: "When compared with these singers, our throats are nothing but clumsy wood. And the sounds that we compose are so bad by comparison that you cannot listen to them; and as long as the great Spirit of Spirits allows us to have these singers, we have the most magnificent music in abundance." The Saturnites who live on the shores of these lakes do not practice any music at all, whereas those who live farther away from these waters and those in the mountains do.

6. Can these birds be captured and tamed? Yes, that can be done. But when such a bird is in captivity, it ceases to sing, even if there is a whole flock. But as soon as it is released and swims again on the surface of the water, the bird begins to sing.

7. These are the singing birds which were mentioned earlier. Someone probably would like to know if these singers are at home in many of the large countries of this planet and whether their habitat is predominantly in the southern, northern, eastern or western parts. This species of birds has its habitat on most of the large mainland or continents. But in the countries themselves they mostly inhabit the southern regions.

8. The northern parts have very few of this species; instead, another species makes its home there, which in a manner replaces this extraordinary company of singers. However, these northern singers of the air do not sing melodies. When several of these birds sing together, it sounds like a wind blowing through the strings of a harp, thus producing enticing sounds. In these regions it rarely happens that these weaker musicians hit a harmonious chord. But for those Saturnites who do not have the opportunity to hear the better singers, the songs of these birds are quite uplifting. Even if they do not sing as beautifully, they are more indigenous. As far as their shape is concerned, they are by far the most beautiful and most magnificent species of bird on this planet. We will discuss their appearance in the next message. But for today, this shall suffice.

9. What do these northern singers of the air look like? It will be very difficult to give you a proper description, since there is not one bird on earth which resembles it. Regardless, we shall attempt to give you an idea as to the shape of this bird. Now listen:

10. This bird is as large as a fully grown ox on earth. Its body is covered throughout with greenish-golden feathers which are more

wooly than smooth. The little feathers on the upper edges of the wings, beginning from the body right towards the ends of the wings, resemble polished gold upon which a bright red carmine color has been applied. The pinion feathers of the wings are light blue; whereas the edges of these feathers look like matte gold. The quills of the feathers are a dazzling white that scintillate in various colors like a gold mother-of-pearl shell on earth. The tail consists of long feathers and is divided into two parts, similar to a swallow on earth. These feathers are not covered with rigid down, but with long and flowing down. The flowing down has colors almost like the down of the peacock's tail feathers. On the outermost tips hangs a bushy mane of these flowing downs, which at times hangs 90 inches long from the feathers. But it is so light that its entire weight in accordance with your measurement does not even weigh 9 grams. These down-manes are multi-colored, so that with every move the colors change.

11. The feet of this bird are completely white and well-shaped, that is, they are not in accordance with the feet of the birds on earth. The difference is that the feet of the earthly birds are usually naked and highly thin, while the feet of the birds on Saturn, right down to the claws, are much meatier and are covered with the most beautiful plumage, which looks exactly like that of the abdominal area, only a little lighter in color. The so-called talons on the feet of the Saturnian birds usually have the shape of the paws of a well-shaped monkey on earth. In this case, however, this bird has feet shaped like a real human hand, only that the fingers up to the pointed nails are covered with beautiful light feathers.

12. This is what this bird looks like from the body to the head, the head being the most peculiar part. Why is that so? Behold, this bird (believe it or not) has two heads, but not in the same fashion as

those earth pictures of a two-headed eagle; instead, these two heads are one on top of the other. A neck like a swan with a head emerges from the crown of the lower head.

13. The lower head is quite round and has a diameter from the bottom to the top of almost 2 feet and is 1-1/2 feet wide. This head has a female human face, almost like the mermaids on earth (which are, of course, very rarely seen), and the top of the head is covered with the richest long hair, changing over into a dark blue color. From the crown of this lower head emerges a 30-inch long neck with a second head which somewhat resembles the head of a swan on earth and which serves this bird in the same manner as the trunk serves the elephant.

14. The bird does not take any nourishment through the second head, because this neck has no throat. Regardless of that, this head does have two eyes and, since it is very flexible, this bird can see everywhere with the eyes of the upper head, whereas it cannot see with the eyes of the lower head. However, since the eyes in the lower head are very sharp, they allow the bird to distinguish everything over the farthest distances. The face of the lower head is not naked; it is also covered with very small pale red feathers; only the lips and the opening of a rather flat nose are bare; everything else is feathered. The eyes of the lower head are large and light blue and towards the forehead the second neck changes over into a dazzling white color. The neck of the upper head is light violet, whereas the head is completely fire-red. The beak is bluish-white with a very strong grip well equipped for holding onto objects.

15. How does this bird take its nourishment and how does it drink? This is done in a very simple manner. It removes the fruit from the tree with its upper head and holds it in front of the mouth of the lower head, which with its sharp teeth, like the mouths of

monkeys on earth, naturally bites into the fruit and consumes it. If this bird wants to drink, it uses the upper head like a drinking glass. It scoops up the water with the upper head and keeps it in the large hollow space of the upper head and then takes it to the mouth of the lower head and the lower mouth drinks from the upper head.

16. That was our second singer, although somewhat imperfect, because it can only sing one tone. But this tone is beautiful and melodious. It would impress the people on earth more than a complete orchestra.

17. And you can believe Me when I say that the music in heaven, when it is most appealing, does not consist of a conflict of many sounds, but instead it is one single tone. This music is the most touching and the most effective. Prove it to yourself. What do you prefer: A single clear beautiful note of a male or female singer, or a piercing chord made by an instrument? When someone has an extremely clear and highly melodious voice, would it not be a pity if such a voice were to be overshadowed by other sounds? Therefore it is not a matter of the multifariousness of the sounds, but it is the quality of the individual tone that causes the heart-rending effect of the music. A perfect tone is within itself already the purest harmony, since a tone does not come individually by itself to a perceptible or audible manifestation; instead, when the fundamental tone comes into being, there already exist within this tone the corresponding tones which are derived from the fundamental tone in the proper proportions, almost like the sound of a pure bell.

18. That is how you must imagine the tone of this Saturnian singer to be, but in a fairly low octave, as for instance the G, A and B which is known on earth as the great octave. This will give you an idea of the singing of this bird. When this bird sings, it begins with pianissimo as much as possible and increases the tone without

deviating in the slightest, or in other words without lowering or raising the tone, to such a decibel level as if you were right next to a bell that is being pealed. The bird keeps this decibel level for a few seconds. Then it lets it fade more and more until you can no longer hear it. Should there be two, three or even four of these birds together, and if they have coincidentally good harmonious voices, this gives a surprisingly wonderful melodious chord, which is enjoyed by the Saturnites.

19. There is, however, no change in this chord; in other words, one and the same chord remains. Even though this kind of music cannot in any way compete with the music of our well-known main singers, this simple music never fails its purpose. Should there be two Saturnites at odds with each other, which only occurs on this planet occasionally, all that is needed is this kind of simple singing, and in no time these two bitter enemies become the most peaceful friends. That is why these birds are often called "peace makers."

20. These birds can also be tamed and then take the same place as peacocks on earth and are considered to be ornamental birds. These tame birds have a stronger voice but their tone is somewhat coarser, while the same species in the wild produces a highly clear and pure sound. The tame species are sometimes taken as a rarity to the southern regions of a country. But there they soon lose their voices as a result of different nourishment; they become sad and sick and usually die soon. That is also why the northern people are not easily convinced to part with any of these birds, because of their fondness for them.

21. The female species bears her young alive and feeds them with very full breasts which are below the neck of the lower head, almost like a human female; only the breast of this bird is not bare, but is covered with light feathers.

22. Now you know everything that you should know about this bird. After this we shall take a look at some of the domesticated fowl, and then we shall turn to the land animals; following that we shall discuss the Saturnites.

22
The Domestic Hen
The Golden Sphere
The Giant Goose
The Nature And Benefits Of These Domesticated Birds

1. As on earth, the hen holds the most favorable role among the domesticated fowl on Saturn. However, the Saturnian hen looks entirely different than an earth hen. On earth, you will find various kinds and species of this fowl in the different countries or regions of the world; this is also the case on Saturn. But there is a bird which is of the same kind wherever you are and which is also known everywhere as the hen.

2. What does this hen look like? It is at least a hundred times larger than a hen on earth. All the hens have the same color. The wings are bright blue, while the back is completely white, changing over to bright red at the tail; the abdominal area has the color of mother-of-pearl, the feet are light red, and the neck is light green from the head right to the area of the feet, which are very close to the head, so that the larger part of the body extends behind the feet.

3. What does this hen look like regarding its shape? Here we experience some difficulty in giving you the proper picture, since on the entire surface of the earth there does not exist one bird with which you could compare this hen. Therefore we have no choice but

to reach for the atypical. Should that be familiar to you, then you will have no difficulty imagining the entire bird.

4. The head of the hen is very large, larger in proportion to the body than the head of a large night owl to its body. On both sides of the head are two white ears, shaped like the ears of an elephant, only they do not hang down. In front of these ears there are two proportionately large, sharp eyes, which are separated by a dark green feathered comb. A little below the eyes is a strong, somewhat blunt gray beak upon which, between the two nostrils, a kind of blood red trunk hangs down, just as is the case with the Indian hens, but this bird keeps its trunk more under a kind of self-willed control. This head is connected to its body by a fairly long but comparatively thick neck.

5. Without the wings and feet, the body has a completely elliptical shape. The wings are proportionately short, and instead of pinion feathers there are only long quills covered with down. However, the part of the wing which is turned towards the head, or in other words the upper edge of the wing, is covered throughout with these kinds of feathers, which are the same as those of the ostrich on earth.

6. As a result of the shortness of the structure of the wings, these birds are not very proficient when it comes to flying. On account of their very long, firm legs, they have the ability to run quickly, and it takes some effort by the Saturnites to capture such a bird. Therefore, if a person wishes to capture a hen, he employs at all times his firm willpower. This willpower will be discussed more in detail at the appropriate time. The tail of this bird is like a wheel, but not in the manner of the Indian chicken, but like the peacocks, though this tail is proportionately larger and more compact.

7. You will have a fairly good idea of the appearance of this bird

if you arrange all the information which has been given in the proper sequence. The only thing left to do is to add to the colors of the feathers a metallic luster, and this will give you a proper picture of this bird.

8. The male differs from the female only in size and often through its annoying-sounding singing, whereas the hen only makes very short broken-off sounds, which are not very pleasant to listen to. The Saturnites quote the following proverb when they hear someone who cannot sing: "Stop singing, you have a voice worse than a hen!"

9. What benefit do the Saturnites derive from this animal? Much the same as the people on earth do from their chickens. These hens lay many very large eggs. The Saturnites drink the contents raw and right away, because that is the way they taste best. The substance which is in these eggs tastes so sweet, similar to good cow's milk on earth, and much better than the milk of the large domesticated cows on Saturn. The shell of this egg is so firm that the Saturnites cut off the smaller end of the egg completely straight. This serves the Saturnites as a drinking utensil, but only for their exquisite juices, of which they say that they are only drunk drop by drop, even though this egg shell, when empty, can easily accommodate 5 pails[22] of liquid.

10. The Saturnites build a living hen house for this domesticated fowl; that is they plant the broad tree, also known as the wall tree, which has the shape of an oval and usually covers an area of approximately 1/2 square geographic miles. In this fairly large hen house all kinds of grasses and plants are grown and, at times, a few of the

---

[22] One pail = 56.6 liters Austrian measurement. This measurement applies here. – ED.

well-known rain trees are also planted. The wealthy Saturnites keep a few thousand of these fowl in such a hen house, as they represent a large amount of wealth to the owner. These birds are compatible only with each other and do not tolerate any other animals in their vicinity; that is why these hen houses are built only for this species of fowl and far enough away from the main dwellings. You can understand why this done if you remember the unpleasant singing of this bird.

11. Besides this bird, there are a number of species which also belong to the category of house fowl; which are not as beneficial as these hens. The meat is eaten, and the feathers are used for beds for the same purpose as on earth. Little is used from the house fowl, because they are mainly for amusement and beautification. Many wealthy inhabitants of Saturn have many different species of tame birds next to their dwellings, whereas others only keep the hens. As far as the remainder of the tame birds are concerned, we will only touch on a few in passing.

12. One of these birds is the "golden sphere." The people like to keep this bird because of the great luster of its feathers, because it is the main attraction or splendor of all their fowl. This bird has the appearance of a ball, which has a diameter of at least 72 feet. Imagine under this ball two strong pillar-like legs with toes that extend in a radial manner. This actually describes the entire shape of this bird; it should be understood that this bird assumes this shape when the wings are folded.

13. The golden sphere has virtually no head; on the front part it has a wide but very short dark red beak which is barely 15 inches long and 120 inches wide. Above the beak there are two oval eyes, the length of which are over 6 feet and the width 4-1/2 feet. The color of the plumage is completely golden yellow; the feet begin

with a greenish color then fade and turn eventually into a reddish color. The entire plumage of the body as well as that of the wings has the same length and is completely flat, without the soft secondary down, and has the luster of a finely polished gold surface. The Saturnites can hardly look at these birds during the day because they have the appearance of a walking gold-plated steeple ball.

14. When this bird dies, nothing but the skin is used. The Saturnites are very proficient in skinning this animal. On festive occasions the skins and feathers serve the ladies as shoulder capes; it looks very good and rich over the shoulder and the full, round arms of these ladies. The eggs of these birds are usually kept for breeding, but usually of twenty eggs there is hardly one which bears a living fruit.

15. These are the most important facts about this gorgeous bird. There is also another which is found quite frequently on Saturn. This bird's body shape is like that of a giant goose. However, that is not its most interesting feature; that distinction belongs to its unusually long neck, which, measured from the body, often has a length of 180 to 240 feet. The color of the body is bluish-gray and the feet are jet black, which is very rare on this planet. The color of the neck is vermilion, but not lusterless, instead it has a very strong metallic luster. The head resembles the head of a goose on earth, but of course in proportion to the rest of the body. This bird body is approximately three times the size of an elephant on earth. The tail does not resemble the tail of a bird; instead, a kind of horsetail dangles at the back part, and its hair reaches a length of 30 feet. As far as the legs are concerned, their length is in proportion to the rest of the body, but they are somewhat on the long side and are as strong as a tree.

16. Those are the features of this bird. Why is this bird kept? As mentioned earlier, usually for the love of splendor. Otherwise this

bird has nothing that the Saturnites can make use of. Occasionally the hair from the tail is gathered and ropes and strings are made from it; however, they are not too firm. The plumage is not used at all.

17. This bird is strictly kept by those inhabitants on Saturn who live on the banks and shores of rivers and lakes. Because this bird is an aquatic bird it nourishes itself mostly on the worms in the water. That is why this animal has such a long neck, with which it can easily reach the bottom, where it finds its nourishment. The difference between the male and female of this species shows itself in its richer tail.

18. This bird lays its eggs in the water and lets them float there for a while, until its instinct tells it that they have completely cooled off. Then it spreads its wings over one or more of its eggs and swims with them towards a quiet place; it keeps close watch over them, and they soon hatch by themselves.

19. When this bird guards its eggs, it is not advisable to come close to such a place, because, with lightning speed, it immediately attacks any intruder with its long neck and deals him a blow with its firm beak, such that no intruder will attempt to disturb this water bird again while it is attending to this most important task.

20. These are the most important and memorable facts of the species of the feathered inhabitants of this planet! You can well imagine that besides the species we have discussed, there are thousands of others in the various countries and continents of Saturn, in the most multifarious variety. Now we shall turn to the tame and wild land animals.

23

The Largest And Most Important Land Animal — The *Mud*
The Teachings Of The Wise About The *Mud*
The Purpose Of Its Creation

1. Because of the many species and kinds of land animals, we shall take a closer look at those which are especially noteworthy and which do not exist anywhere else except on this planet.

2. The largest living land animal on Saturn is called the *mud*. It can only be found on a very few continents on Saturn. If you would add them all up, you would not find more than 10,000 such animals on this entire large planet. The countries where they can be found are sparsely populated. Because of the *mud's* large size and tremendous voracity there is not much left for any other creature. Also, the Saturnites do not have the courage to fight with this giant animal. They relinquish the land which is occupied by these animals without hesitation and they call it the "uninhabitable *mud*-land." This animal does not exist on the main continents. But south and north of these main continents there are other large islands, and these countries are mostly left to all kinds of those species and other animals. But none of these countries are avoided as much as those that are considered to be *mud*-land.

3. What does this animal look like? Does a similar animal exist on earth? Yes, there is a similar animal on earth, but this animal plays very much a subordinate role, while on Saturn it occupies the foremost terrible rank in every aspect; this applies to its giant size as well as its ferocity and voracity.

4. Which earthly animal does this creature resemble? It is one which is well-known to you: it is the pig. But as far as its size is concerned, the pig on earth would hardly be large enough to be a

parasite by comparison. Even these huge Saturnites feel no larger than tiny little dwarves at the sight of such a giant animal. I tell you, if the *mud* would stand behind a high alp, for instance the Chor Alps,[23] you would still have to stretch your neck in order to see the apex of its back.

5. This immensely large animal is just as voracious as your little pig, as it is not very particular about its food; it eats whatever comes along, may that be grass or certain trees, or other species of animals, human beings as well as water animals — it eats them all with the same appetite.

6. Since this animal has power in accordance with its size, it would be vain to enter into any kind of combat with it. Some courageous Saturnites have made the attempt with long sharp tips which they mounted onto 600 foot-long sticks to control such animals and in addition they employed their total willpower. Despite this, they were badly injured. The *mud* was wounded in many different areas and, because of the pain caused by these wounds, became very angry and ferocious and ran to a wide river in order to cool its wounds. As soon as the pain eased, the *mud* got up and scooped up a huge amount of water as well as some large stones from the river bed and spit them on its pursuers, who were under the impression that they had conquered this animal; they were all badly injured and only a few made it back home. Some of those that were killed by this animal were devoured with only a few bites as soon as the animal reached their location on land.

---

[23] 2,000 meters high. – ED.

7. In order that you may have some idea of how large the *mud's* mouth is, I shall give you an example: if there were nuts larger in size than your Schloßberg,[24] such a nut would not be too big for one tooth of this animal to crush easily with one bite. If the *mud* were to fill its mouth with water and stones and spit out its contents from the upper part of Styria to the lower, the whole region would be flooded and the waves would reach higher than the highest steeples of your city of Graz.

8. If you take this information into consideration, this will answer your question as to why the Saturnites do not entertain another encounter with this animal. This is also why the Saturnites have only made three such attempts since time immemorial, and each was a failure. And for now, they have given up this initiative. Their sages have the following advice:

9. "A human being is capable of doing many things with his power, but he cannot control the moons, the large bright ring, the streams, the storms on the ocean, the giant fish and the *mud*. Therefore he should use his power where it can be effective. He should not attempt to control superior things with his limited powers."

10. And there is also another advice or teaching of the sages which is as follows: "Listen! The Great Spirit gave us a great world to live in; we do not know where it begins or where it ends. In the land where we were born we know of the things in the water, on land and in the air, and we know and have experienced at all times that they were subject to our powers. If it would have been the Great

[24] A mountain in Graz (475 meters in height). – ED.

Spirit's intention that the *mud* was meant for us, then it would also obey our powers. We know, however, that it scoffed at our powers with the greatest ease when we attempted to make it subject to our willpower. Therefore it is as bright and clear as the sun that shines throughout the day that the Great Spirit has appointed other powers besides ours that should not serve our powers; and we should not make those powers our servants. That is why we should remain within the borders of our powers as they have been determined for us, and let other great powers work where the Great Spirit has placed them. It should be of no interest to you to know what the great bright ring and the moons above us are. And we should never, now or in the future, set foot on *mud*-land!"

11. If you think about these words of wisdom for a short while, the purpose for this animal's giant size and tremendous strength will probably become clear to you. It would serve no purpose to describe the shape of this animal any further. It is left to the imagination of the individual to imagine this animal as well as he possibly can.

12. Do the Saturnites at times keep an eye on this animal? That happens rarely. This only happens on the occasion of a long voyage, or when it is seen from the foothills of a main continent where such a *mud*-land is not too far distant. It is not advisable to get close to the shores of a *mud*-land. If this animal notices anything from its vantage point on land floating on the water's surface, it will walk with giant steps into the ocean, and if the ocean is not too deep it will succeed, with only a few steps, in catching whatever is floating on the surface and welcoming it with its immense mouth.

13. Something that is especially frightening and horrible is the grunting of this animal; truly, you have no idea how horrible it is. I can tell you this much: if this animal were as far away in Hungary as it could possibly be, pointed its mouth towards Styria in Austria

and grunted a few times, not only would the earth tremble in Styria but much farther. Not only would every building in Graz be leveled by this type of earthquake, but also the loose mountain tops of the neighboring Alps would tumble. This description should give you an answer as to why the Saturnites are not the greatest of friends of the "moving song" of this animal.

14. This animal has very keen senses despite its tremendous size. Its hearing and smell are especially keen; that is why it can perceive something suitable for its stomach coming closer on the water surface even from the farthest distance. This pig is by far not as unclean as the pig on earth. Especially regarding defecation, the *mud* surpasses in cleanliness almost any animal known to you on earth. Before it releases its bodily waste, it digs a very deep hole in the soil which, in accordance with your measurements, has a circumference of about an hour's journey and is about 600 feet deep. Once such a hole has been dug, the *mud* disposes of its waste in this hole and covers it immediately with the same soil that it dug out. That is how this animal keeps the land it occupies clean and fertilizes it accordingly for the growth of grass that follows, which on these *mud*-lands is usually by far of a greater density and larger than the densest jungles on earth.

15. What remains to be answered is this: What is the purpose of such a colossal animal on this or any other planet? The answer was given in the description of the giant fish. As this fish forms the general transition from an aquatic animal to an animal of the air, the mud is also such an animal; it forms the transition of all levels of animal and vegetation into a nobler species of animal which is closer to a human being. Now you know everything there is to know about this animal. In the next few chapters we shall extend our view of the land animals, but with greater expediency.

24
The Saturnian Elephant, *Sisterkihi*
Its Structure And Habits
Hunting This Animal And Its Purpose Of Creation

1. The animal which we shall now describe is called the *sisterkihi*. To what order does it belong on Saturn? This animal is the actual elephant of this celestial body; despite this, it is far behind the *mud* in every respect. In size the *mud* is more than a hundred times larger than the *sisterkihi*; it also does not resemble the *mud* at all. The *sisterkihi* greatly resembles the elephant of earth, but not as much as many of the other animals which we shall hear about.

2. What does this animal look like? It has four extremely strong feet, similar to those of the earthly elephant, but in proportion to its size. The legs are not structured like those of an elephant, but more like those of a bear, with strong claws. Its body is extraordinarily voluminous, to the extent that when the *sisterkihi* is fully grown it measures 420 to 480 feet from the abdominal area to the spinal column. Its tail is proportionately as long as the tail of a lion. At the end of the tail is an extremely strong bushy tuft of which the length of the hair is anywhere from 36 to 60 feet. This animal has a proportionately very long but massive neck, with a comb that extends to the area of the front feet consisting of a very strong hirsute mane.

3. The head is similar to that of a horse; in proportion its forehead is much wider. Above the forehead, between the two ears, there is a large flexible trunk which expands to 240 feet, whereas on the other hand it retracts to one-third of its full length. The trunk has a tremendous lifting power; it can uproot trees of a large size.

4. By means of this trunk this animal tears branches off the trees

and devours them fruit and all, especially when it is very hungry. The color of the *sisterkibi* is light gray while its trunk is dark gray. Its neck-mane scintillates into a blue and the tuft is light gray. Its eyes are of a very dark brown color. This animal has an abyss of a mouth with extremely strong white fangs. In the back of its mouth it also has extremely strong bones that are used for crushing its food; they are not teeth, but are actual bones of the upper and lower jaw protruding prominently. With these protruding bones this animal can even crush very hard stones as well as big branches with ease, to the point that they turn to mush in its mouth.

5. This animal is actually a herbivore; however, when it is very hungry it will devour animals, and not even human beings are safe. This is one of the reasons why the Saturnites are not very fond of this animal. But wherever the Saturnites may find this animal, they have the ability to deal with it, even though it might present some great and dangerous obstacles to them.

6. How do the Saturnites hunt the *sisterkibi*? They cannot hunt it with mere strength, because this animal has as much strength in its trunk alone as a thousand of the strongest Saturnites. That is why the Saturnites choose shrewdness as their ally. This animal's habitat is usually the large forests of the well-known pyramid trees, and it causes great harm to these trees by tearing off the branches as far as its trunk can reach.

7. When the Saturnites notice in any of these countries that these giant trees are without any branches (and, by the way, these trees are native to almost every country on Saturn), it means that one or more of these animals are present in the forest. What will the Saturnites do in such a situation? Equipped with torches, they completely surround the forest and ignite it all around. Since this animal fears nothing more than fire and smoke, it soon flees from

the forest and searches all over for an exit where the fire has not spread yet; but wherever they find such an exit, the Saturnites make certain that it will lead to a wide river, a lake or even the ocean itself. When the *sisterkihi* finds such an exit, it will soon enter the water.

8. As soon as it enters the water it becomes awkward and clumsy and holds its otherwise active trunk straight up in the air. The Saturnites now force the animal to go deeper and deeper into the water by pursuing it with their boats and torches attached to long poles. Once the animal has reached a depth where only its head is above the surface of the water, they make haste and hack off the formidable trunk with large sharp axes. As soon as this animal loses its trunk, it loses its life. It drowns and begins to rot, and its flesh becomes nourishment for many hungry animals in the water.

9. What happens, if for one reason or another, such a forest does not border any water of the required depth? Then an operation like this becomes somewhat more dangerous and precarious. Should the fire spread properly from all sides so that this animal or several of them are surprised in the center of a burning forest and cannot find an exit anywhere, they will suffocate in a raving frenzy. If the fire is very severe they will also burn. However, should the fire not have taken proper root, they will break through with great ferocity towards an area where there is less fire.

10. Beware to those who encounter this animal under those circumstances! It will pick up any human being or animal with its trunk and hurl it with such vehemence into the air or onto the ground that nothing remains. The force of such a frenzied thrust surpasses the force of a cannon ball. If this animal would carry out such a throw on earth, it could easily throw an 11,000 pound stone over a distance of 20 geographic miles, but with such force that the

stone would travel this distance in just a few seconds. Should this animal hurl it into the ground, then you can be assured that the stone would be driven over 600 feet into ground.

11. From this brief description of these circumstances you can easily surmise the respect the Saturnites have regarding this animal and the concern it causes on entire continents when one or several of these animals have escaped such a fire hunt. That is why such a forest is at all times subject to severe scrutiny; it does not matter if it is located near a body of water or not. Should the forest not border on any body of water, then there will be many deliberations on the subject of if and when it would be most favorable to burn the forest.

12. Should the forest be in a good burnable condition and if at the same time it is large enough, then it will be burned. Should this not be the case, then the forest is for the time being relinquished to the animal, but in the meantime dry wood is placed around the entire forest, literally forming a wall around it, which is ignited when the trees become more burnable. This usually occurs when an area comes under the shadow of the ring for extended periods of time, which in accordance with the chronology on earth extends over several years. During this time the trees stand just as on earth in the winter, without sap. During this time the Saturnites ignite this wall of dry wood from all sides at the same time, then leave as quickly as possible.

13. This concludes the chapter on this memorable animal. The *sisterkihi* inhabits approximately ten continents on Saturn, but in small numbers.

14. Regarding the benefits of this animal, in a psychic or astral respect the same circumstances exist, but to a somewhat lesser

extent than with the *mud* and the giant fish.[25] This animal, as well as all the others, also has another natural purpose. And none of these animals have been placed upon this planet as an indispensable prerequisite for the maintenance of other entities, as these transitions can pass just as well through other levels. For this reason no country is at a loss when such large, strong animals which are dangerous to human beings become extinct.

25
The Blue Bear, *Ihur*
Its Structure, Character And Nourishment
Its Usefulness As A Cultivator Of The Soil

1. After we have learned about these two giant animals of this large planet, we shall turn our attention to a few other animals, which are not as enormous. They are, however, important enough to deserve some thought.
2. The first animal in this order is *ihur*; in your language it translates into "blue bear." When the *ihur* is fully grown it is almost as tall as a human being on Saturn, that is, if this animal stands upright on its hind legs, which they usually do. The name of the animal discloses its color, which is light blue throughout.
3. What does this animal otherwise look like? In general it looks like a golden bear on earth, but the structure of its head is entirely different.

---

[25] That is, for the development of the souls in nature. For details read Chapter 17, paragraph 7 and Chapter 23, paragraph 15. – ED.

4. What does the head look like? It will be a little difficult to describe it in order that you may get a proper picture of this animal, because you do not know any animals on earth which have a head similar to that of the Saturnian bear. Nevertheless, we shall describe a shape by which you can view its head.

5. Imagine a fairly round skein of wool having a diameter of 9 feet, and on both sides, approximately in the middle, there are two long ears which have a length of 9 feet and a width of 6 feet. Imagine furthermore two long winding horns at the uppermost part of this skein, each being 18 feet long and approximately 3 feet apart from each other. The color of these horns is a lusterless gold, and approximately 5 feet below them are two proportionately large eyes which have exactly the appearance of a human eye. Below the eyes imagine a proportionately large mouth like that of a lion. Furthermore, this head is connected to its body by a proportionately massive, long, strong neck.

6. In addition behind its horns on both sides of the neck this bear has a 12 to 18 foot long mane of a darker blue color; there you have the entire shape of this animal. The *ihur* has no tail, however in its place there is a little longer and darker hair.

7. If you compose everything that has been given in the proper manner, you will be able to imagine this animal if you add to this that the *ihur* often measures several times 300 feet beginning with the horns and ending with the hind legs, and when this animal stands on its four feet it measures almost 120 feet from the ground to the upper back, and each foot has a length of 36 feet and a thickness of a barrel that surpasses a volume of 566 liters. The paws are like that of a bear; they have the same shape, only the size and color are in accordance with the size and color described here.

8. Hopefully, it will not be necessary to describe this animal's size and shape in any more detail. Let us therefore describe its character, its mode of life and its usefulness.

9. The *ihur* is usually good-natured; however, it should not be provoked or pursued. If provoked, it soon changes its good-natured character and becomes quite vicious and enraged; while in this condition nothing is safe. Whatever crosses its path will be immediately attacked and totally destroyed. Although this animal is not any taller than a human being on Saturn, it has the strength of ten men; that is the reason why a malicious person on Saturn would not fare well if he came in conflict with such an incensed animal by himself.

10. The Saturnites shy away from this animal despite its good nature; they try to chase it away with all kinds of means and drive it from areas populated by human beings. That is why this animal is very rarely seen by anyone.

11. What is the nourishment of this animal? It nourishes itself with grass, roots and young branches from trees and shrubs. It never eats flesh, not even in an emergency. If provoked, it tears apart humans and animals, then lets them lie there and leaves.

12. A very peculiar characteristic about this Saturnian bear is that it has a very peculiar fear of its own anger. For this reason this animal, guided by its own instinct, carefully avoids, wherever possible, any situation where it might become enraged. Such an instinct would be a blessing for many human beings on earth, especially those who are ambitious quarrelers and belligerent patrons, who do their utmost to find situations where they can quarrel. This information should suffice. We shall now turn to the usefulness of this animal.

13. The *ihur* can rightfully be considered the cultivator of wilderness areas; in a very short time it loosens up large areas of soil with

its extraordinarily strong claws, and it does this so well that even the Saturnites with their good tools would not be able to match it. What does this animal do after it has loosened the soil? It looks for the fruitful areas and searches for all kinds of edible root vegetables and places them into the loosened furrows. The intent, of course, is not to cultivate a field, but rather to gather a supply of food. But it happens that some of these roots that are placed in these furrows are not eaten by the bear and then they begin to sprout and grow. In this manner a field that is wild and barren becomes fruitful, and even more so when this animal has filled its storage abundantly, because under those circumstances it will not leave until the supply is almost consumed.

14. Because it has walked about in an area, as long as it has found something to eat it leaves its droppings everywhere, and by so doing it unintentionally fertilizes this area for years.

15. When human beings on their frequent travels in these large countries find such places, they know that they are in the neighborhood of such an animal. They wait for extended periods of time to find out if this animal will make further use of such a field. Should this not be the case, that is proof to them that the ihur has left this place; and then they will take possession of this field.

16. It does happen occasionally that the bear returns because of the roots; then the inhabitants either do nothing and allow the bear to dig its furrows again and, in so doing, destroy the crop they have planted, or else they have to attack the unwelcome guest with force; such a confrontation always has a dubious outcome. The animal will always fight for its prerogative and will not easily leave. As for the human beings, it is not easy for them to leave such a new fertile land either.

17. Should such an animal be conquered or driven out, then the

inhabitants do not have to fear any further disturbance or its return. Even if they are not able to kill this animal, by pursuing it they get the animal to the point that it notices it has been hunted. And since this animal, when in a quiet state, fears its own anger, it will not return to the place where it has been provoked.

18. This is everything which is noteworthy about this animal; we shall now continue with another animal that inhabits this planet.

26

The Saturnian Lion, *Horud*, Serves As A Hunter And Lumberjack

Method Of Catching The Young Saturnian Lions

1. The name of this animal is *horud*, which we shall now discuss briefly. Where does this animal belong in the order of animals on Saturn? If you observe the lion on earth, the *horud* has the same position on Saturn. But does this animal look like the earthly lion? The answer to this question is neither yes nor no. In many ways it resembles the lion of earth, but there are also many aspects that are not the same. A closer description will point out the differences in its shape from the earthly lion.

2. What does this animal look like? It is just as large as the blue bear, but it has more than one color, in accordance with its body parts. The back of this animal is bright red, extending close to the middle of the abdominal area. The shoulder blades, the front legs and hind legs are pale green. The abdominal area itself is more of a dark green or a lush grass-green. The tail is white, whereas the bushy tuft at its end is light red. The upper side of the white tail is adorned with red spots in regular intervals. The claws on the feet are also white, and the backs of the claws are trimmed with a red stripe.

3. Except for the different colors we have just described, and the neck and head, which do not resemble those of the lion on earth, the rest of the shape of the *borud* completely resembles that of an earthly lion. There is not an animal on earth which has a head resembling that of the *borud*. This animal has an almost square head, just like a die where the corners are rounded off. This head sits on the neck in such a way that the rear surface is attached to the neck, but not the front part because the front surface projects from the neck like a jaw by 3/4 of its diameter. On both lateral surfaces of this head are two crescent-shaped ear funnels which are located backwards from the surface on each side of the head by over 6 feet and have the colors of the rainbow but in very light shades.

4. On the upper surface you find an almost 3-foot long, conical and completely black horn. That is the basic color of this horn; but on this black surface are round reddish disks with regular snail-like spirals which have a strong metallic luster. Around the bottom of the horn is a wreath of long light blue hair, which gives the horn the appearance of sitting in a basin. Towards the back of the head and the back part that connects to the neck, the hair becomes continuously longer and denser, whereas towards the forehead the hair becomes shorter and curlier.

5. On the front surface of the head or face there are, in proportion to the animal, two very large eyes located in a cavity, each eye having a diameter of 3 feet in accordance with the measurements of earth — that means of course the actual eye itself. If the cavity and the corner of the eye would be included it would be almost 6 feet in diameter. The eyelids are a very dark red, above which are proportionately very large thick eyebrows, similar to those on human beings; they are curly, just as the hair around the horn towards the forehead.

6. Now we have come to the most peculiar feature of this animal, which is its mouth. At one time or another you must have heard about a bird called the griffin. That is what this animal looks like, with the exception of the wings. Instead of the usual mouth it has an enormously strong eagle's beak, which is similar in color to the horn on its head. The spots or disks are not snail-like, but in the form of rows from the base to the tip of the beak and decreasing in size. The upper part of the beak is like any other bird which you know on earth. The upper part does not move, whereas the lower part of the beak, inclusive of the lower surface of the head, is movable right to the neck. At the base of the beak, this animal has mighty crushing teeth. Instead of the incisors or fangs it uses its very mighty and strong beak which protrudes by almost 9 feet from the face, and at the base it is almost as wide as the main surface of the face.

7. The *horud* also has a tongue that can be remarkably extended in length, and it has almost the attributes of a trunk. Therefore this animal has the capability to seize all kinds of things by force with its tongue and draw them into its mouth. Around the base of the beak there is also curly light blue hair that turns greenish towards the neck.

8. What is usually the color of the head? The head is light ash in color; under the eyes as well as on the forehead there are three bright red circles one on top of the other. As far as the other parts are concerned, only the two lateral surfaces with the ears are visible and are also of an ash gray color, but are plain without ornamentation. As you know, the back surface, beginning with the upper surface, is adorned with long hair, and the blue color becomes livelier the closer it comes to the neck. The neck is fairly massive and its length to the head is just as long as the entire back part (that is,

beginning from the shoulder blades, the front legs to the tail) and covered throughout abundantly with manes of a bright blue color. That is what this animal looks like.

9. What is its usefulness? What kind of character does it have? Where is it at home? And what is its relationship to the Saturnites? We shall briefly answer these four questions. The *horud* is usually of a gentle nature, and is frequently kept by the Saturnites as a tame animal. They enjoy this animal's splendor and beauty as well as its willingness to work, provided it has been properly trained.

10. What kind of work is this animal used for? Usually to hunt several other kinds of animals, which are of the kind that do little harm and at times great harm. This animal is also employed as a lumberjack during the time of the shadow; with its beak it bites off thick branches from the pyramid tree with one bite, and it can climb this tree to the top with the greatest ease. The branches often have such a circumference that on earth it would take five men to encircle them. It does not take any more effort for this animal to bite through such a branch than it does for you to bite into a tender apple.

11. Once the *horud* has bitten off enough branches, it will then pull these with its beak to the dwellings of human beings after being given the proper signs to do so, and while there it will bite off smaller pieces as indicated by its owner. These pieces of wood are then used by the Saturnites as firewood. They also employ this animal for many other kinds of work such as biting and carrying.

12. The only prerequisite is that this animal must be captured when young in order to be trained. Because if a mature animal is captured, it will not become accustomed to this kind of work, and as such would be of little benefit. Besides, when it is not provoked the animal will run away as soon as a human being comes close, and

when it is surrounded by human beings it cannot be trusted for the longest time. Because as soon as it begins to dig into the ground with its strong claws, that is a sign for the hunters that it is high time to leave. If the hunters do not leave, this animal will begin to jump and roar, and whomever it strikes will discover the power of its beak and will feel as if he has been hit by a big tree trunk. That is why the hunters retreat when this animal begins to behave in this manner.

13. If the *horud* is this dangerous, how are its young captured? The Saturnites use their cunning. Because the people who live in the regions where this animal is at home know that it is very partial to intoxicating alcoholic beverages, but only at those times when they have young; this applies to the male as well as the female, which are the same in appearance except for their genitals. When this situation arises, the hunters on Saturn carry fairly large containers of such beverages into the vicinity of these animals. And it does not take long until they take the bait. As soon as the *horud* has emptied the container, it very gently returns to the place where its young are, and there are usually two, three or four. Once it reaches the spot it falls into such a sound sleep that it does not even notice when its young are taken. The young are taken to the homes of the Saturnites and are trained for their particular purposes. The mature animals are left alive for further propagation.

14. That is everything you need to know about this particular animal; the only thing which you do not know is where its home is. Its habitat is only in the southern regions of Saturn and only on those continental countries which do not extend beyond a 45° southern latitude. Since this animal loves the ocean regions, it is only at home where it can enjoy the necessary temperature it requires. Should a country exceed this degree significantly, naturally it will

get colder, especially by an ocean, and climatic conditions like that would of course be completely unsuitable for this animal.

15. The *horud* has this special instinct in that it will not inhabit the western or the eastern parts of a country; it will only inhabit the southern center. Should this area be large enough, then it will live in such a country. If this type of environment does not exist, then it cannot progress and cannot therefore maintain itself. That is why the *horud* will not be found inland, but always in the regions of the oceans. Sometimes, however, it is taken there as a curiosity, but it will not survive for long under those circumstances.

16. Now you know everything about this animal which is of importance. And now we shall turn to one which can be found only on this planet.

27

The Saturnian Antelope, *Zigst*
Its Purpose And Creation Hints
The Reasons For Hunting *Zigst*
Fraud With Secret Remedy

1. The animal which we shall discuss next is named *zigst*, or, in the language of earth, "pointed foot" or "piercing foot," and it is native only to this planet. In the order of animals it is approximately on the same level as the antelope on earth. Its habitat on Saturn is only on the highest mountains.

2. Why is it called "pointed foot"? Do not think that this animal has four pointed spears in place of the actual jointed feet. It is called pointed foot because on its front feet, in the area of the usual members above the claws, it has no member; instead it has a straight horn which is sharply pointed downwards. At the same

time, this claw extends from the area of the knee as a completely solid claw. The hind legs of this animal are like the legs of any other animal; however, the claws are not split, but are very pointed.

3. These are the reasons for this animal's name. What does this animal otherwise look like? Among the larger animals on earth there is none which is like this one; there are some similarities, however, among the smaller animals. The main body of this animal resembles the body of an otter; the tail resembles the tail of an ox. The neck and head have a close resemblance to a tiger; but the teeth resemble those of a graminivorous animal.

4. At the crown of its head it has a single horn which is somewhat bent back. This is a description of this animal's shape, other than its size and color.

5. What is the size of this animal? If you were to measure its size in accordance with earthly measurements you would not find one animal on earth with which this "pointed foot" would compare. Because on Saturn the ratio of everything is at least a hundred times larger and at times even more. This "pointed foot" belongs to the smaller animals, and barely measures a third in size of the previous animal which we have come to know as the lion of this planet. This is also the reason why it is easy for every human being on Saturn, after the *zigst* has been caught, to carry it home on one's back.

6. What are the colors of this animal? The main color is a dazzling white, but from the head to the tail it has a wide light-blue stripe. Towards the abdomen this animal is golden yellow, while at the feet the color changes to reddish, except for the pointed claws which are completely black as well as the horn on the head. The underside of the neck, starting at the lower jaw right to the chest, is striped dark red.

7. Now you have the complete shape of this animal, which

cannot be found in this manner on any other planet. What is this animal's usefulness? What is its nourishment? Do the Saturnites capture many of these animals?

8. As far as its usefulness is concerned it is of no more importance to the Saturnites than a chamois or an antelope to the people on earth. But nevertheless it has its place in the order of things, which it unknowingly occupies usefully. Who knows the usefulness of the chamois on earth? Who can give you the reason why this animal climbs the mountains? But to whomever wishes to know, I shall disclose the reason.

9. You are aware that all kinds of mosses as well as species of plants grow on the high mountains of earth for the purpose of dissolving the rocks. You also know that moss and plants are products of spiritual potencies and spiritual intelligences. It becomes abundantly clear that they are such products when, within them, some kind of intelligent life has begun to express itself. Once life begins to express itself, it does not do so to relapse again into death, but instead to strengthen itself by developing into a form, in order to leave that form and pass over into a higher one.

10. Which life-expressing form in the Alps or mountains are above these small living forms of moss, grass and all other alpine or mountain plants? These are of course the animals of the mountains. They are the higher life forms into which the plant life of the high mountain regions passes over.

11. You can easily understand that this is the proper sequence of events, because the life of the animals is maintained by incorporating the life of the plants. And that is why nourishing oneself from food which is compatible with the nature of the animal is nothing more but to absorb and unite the scattered life of the lower smaller potencies to a higher more complete life. In other words:

12. To nourish oneself means to gather and absorb the life in a container that continuously emanates from Me, so that it becomes stronger and more complete from level to level on its return to the primary source from whence it came.

13. If you comprehend what has just been said to some degree, you can then apply this concept to the *zigst*. Transfer the same quality to the "pointed foot" and you will know everything there is to know about this particular aspect.

14. There is another question which needs to be answered: namely, whether the Saturnites capture this animal. My answer to this question is that at times the very courageous Saturnites often go out and attempt to capture this animal, but very rarely are they successful. Because this animal is so skillful in climbing the highest mountain peaks of this planet that none of the human beings are able to follow it. This animal also has the ability, on account of its pointed claws, to stand on an area no larger than the palm of a hand. Where the rocks turn into rugged peaks, there no longer exists any possibility for the huge Saturn beings to continue their hunt.

15. When such an animal tumbles down from a high and steep mountain peak and makes such a fateful fall, which happens very rarely, this is the only time that it can be captured, if this place is accessible to a human being. So far, not one human being on Saturn has captured the "pointed foot."

16. You might question that if this animal is so hard to capture, why do the Saturnites make the effort? They are compelled to do so by a kind of superstition. On earth this type of superstition falls into the category of so-called medicinal quackery. Because this animal nourishes itself from the most powerful and most fragrant herbs, the Saturnites are of the opinion that its meat must be so very healthy

that the one who eats it, even when it is very little, will become immortal.

17. There is not much difference between the Saturnites, who know all kinds of remedies through which they believe they can perpetuate their lives, and the people on earth when it comes to this type of medical belief. However, experience teaches us on a daily basis that the death of the body cannot be halted.

18. What do these people do, in spite of their daily experiences which prove their remedies continuously without effect? They place the remedy into an extraordinarily secretive category of their restorative science by saying: This remedy must administered exactly at midnight and in an exact dosage. One thousandth gran[26] more or less makes the remedy ineffective.

19. Should this explanation not suffice, they confuse the matter further and seek refuge in the influence of the constellations. Then these mystical "life-extending medical practitioners" advise their clientele with great and highly unintelligible eloquence as to which quarter the moon must be in and into which constellation the sun must pass, and they say that this has to occur exactly at midnight. If, for example, the Sun is not in the sign of Leo at midnight and the Moon is not in Capricorn at the same time, and if a certain planet does not pass into a certain sign and another planet into another sign of the zodiac at the same time, then the "eternal life remedy" is powerless and ineffective.

20. Gullible people usually believe in such mystical preachers of wisdom or wiseacres, and buy costly remedies which are heralded as

---

[26] 1 gran = 2 drops of water or the size of one peppercorn. – ED.

providing eternal life. Once these people are in possession of such an "eternal remedy," they look in all kinds of calendars or ephemerides to find out when the Moon, the Sun and the rest of the planets enter into the predetermined signs exactly at midnight. You can see that even without any particular knowledge in mathematics, these astronomical and astrological transitions and relationships as to the signs of the zodiac will *never* occur; and if they should occur, it might happen once in a million or several million years, and not exactly in the manner that the remedy prescribes; therefore the effectiveness of such an eternal life-bringing remedy completely neutralizes itself in accordance with this mystically cunning speculation. Under these circumstances the seller of these remedies escapes liability, because he can make the claim that the remedy was not taken in accordance with the instructions.

21. On Saturn the meat of the *zigst* is used in a similar manner for a similar purpose. However, when the remedy does not produce the desired effects, then the life-prolonging doctors on Saturn use the excuse that the person using the remedy was somewhat careless and, instead of taking it during the time of the shadow of the ring, must have taken it during the sunlight; and under such circumstances the remedy is ineffective.

22. Should a relative of a departed person complain to such a life-doctor and mention the fact that the departed did take the remedy at the time of the shadow of the ring, then the doctor will question the relative as to the position of the moons. Should the relative be able to provide this information, then the position of the moons comes into question and is considered by the doctor as highly unfavorable for the remedy, and he explains why this is so with the greatest eloquence. Most people cannot answer questions as to the position of the moons or planets; this then becomes an escape clause for the "eternal life bringer."

23. It may also happen that a relative of a departed person who has taken such life-bringing remedy consults yet another of these "life-bringing doctors," regarding the fact that the remedy was ineffective. In a situation like that, you can well imagine what this life-bringing doctor will say about the useless remedy of his colleague: "Why did you not come to me? It is well-known that this doctor uses the wrong kinds of remedies!" And in order to convince this person that the deceased must have been given the wrong remedy, he shows him a remedy that has a different color. That usually convinces the person making the inquiries as to why the other remedy was ineffective.

24. In these situations the relative of the departed mostly goes back to the original doctor, whom he now considers to be a charlatan. How does such a doctor get out of a situation like that? The doctor takes the person who demands accountability to neighbors minded like himself and likewise informed, and says to the person who demands accountability: "See, these people have used my remedy properly. Ask them how old they are!" Should he then question them as to their age, he will usually get an answer which far exceeds even the age of a person that has achieved a very old age. Usually this is done by not mentioning the number of years, but expressing things through extraordinary facts, such as experiences which they had during their long life, which are to be proof. For example, such a person might mention that he remembers well that a certain mountain did not exist at a given time. Another one might say, by pointing to a light white stripe on the firmament, that he saw when the Great Spirit spanned that ring across the firmament. Another one might say that he remembers the time when there was no moon on the firmament. Each person who is questioned about his age has a better reason to prove his age than did his predecessor.

When the person who demanded proof hears these testimonials he is usually satisfied with these explanations and might, in addition, even buy another remedy from such a doctor, who is not any older than his neighbors.

25. This is all the information which is necessary to know about this particular animal. Let us now turn to another animal on this celestial body which is not tame, and then we will begin with the domestic animals.

28
The One-Eyed *Bauor*
Hunting The Animal
The Skin As A Coat For The Patriarchs

1. The animal next in line and worthy of mention is very rare. In the countries of the continents it is a complete stranger. It is only native to some of the important southern islands. Whatever island this animal calls home, the Saturnites visit these islands as seldom as they visit the countries where the *mud* lives. Why this is so will be explained in the chapters that follow.

2. This animal's name is *bauor;* in your language it means as much as "one eye." First we shall discuss why this animal is called "one eye." Does it actually have only one eye? No, it has two eyes to see with, like any other animal. But above its two eyes, in the center of its broad forehead, it has its weapon eye. And the animal's name is derived from this dangerous eye.

3. But before we describe this eye, we shall describe the animal in accordance with its form and then we shall describe this eye.

4. What does the *bauor* look like? What is its size and color? It resembles a horse with the exception of the neck and tail; the body is of course a hundred times larger.

5. The tail resembles a snake without a head, and it reaches lengths of 720 to 780 feet, and in the anal area where this tail has its beginning it has a diameter of 9 feet. At the end of the tail it has three very strong barbed hooks, similar to a ship's anchor. This animal has most of its strength in its tail; with it, the *bauor* searches for its nourishment in the water; this is why it can usually be found at the shores of the ocean, with its long tail constantly moving about the water in order to catch a large fish or some other large water animal for its nourishment. The *bauor* is also very skillful in this matter. As soon as it suspects something alive in the water which could be for its benefit, it shoots its tail like an arrow to that particular spot, and when it projects its tail in this manner it never fails its target. This is the purpose of its tail.

6. What does the *bauor's* head look like? Its head is extremely peculiar, as it sits on a long, strong neck and almost resembles the head of a sea-calf or common seal on earth. Only the head is proportionately larger than the head of a seal on earth, the same as the body is larger than that of an earthly horse. Except for the weapon eye, the head resembles the head of a seal!

7. What is the actual purpose of this weapon eye? It is actually not an eye to see with, but an eye to feel with and to transfix or capture with. Usually this eye is closed, but should an enemy or something unfriendly move into the vicinity of the *bauor*, then the eye will open. And as soon as this eye opens, an intensive red ray of light bursts forth and it would be much easier to look into the noon sun than into this eye.

8. When this ray is directed upon any living being, it is soon captured to such a degree by a kind of heaviness that it feels spellbound and is unable to move from the spot where it was struck by the ray. When the *bauor* notices that its enemy is sufficiently unable

to move or spellbound, it slowly closes in, which is at approximately the distance of the length of its mighty tail. This animal never makes mistakes as far as the distance between its enemy and the length of its tail is concerned. Once that has been established, it projects its tail as quick as lightning upon its enemy and picks it up; the tail is used as an arm, and it places its catch into its extremely large mouth and crushes it with its powerful teeth for nourishment. This animal makes no distinction between an animal or a human being; because of its great voracity it eats any living being, whether it be an inhabitant of the air, earth or water.

9. Except for the color of this animal, that is everything noteworthy about it.

10. What is its color? In the abdominal area it is light blue; along the center of the abdominal area passes lengthwise a considerably wide dark stripe. The back of this animal is light red with small zebra-like yellow stripes. Beginning at the body, the legs are an orange-yellow. The hooves are completely black. The hair on its body is fairly short throughout; that is why the Saturnites are of the opinion that this animal was completely naked. But when a few of these animals were caught, they noticed their error.

11. If the Saturnites wish to capture the *bauor*, it becomes a very expensive proposition for them, because if they want to apprehend it they must sacrifice many of their domesticated animals beforehand. Not until the *bauor* is surfeited with eating many of these animals does it become tired and exhausted; it lies down somewhere close to the shore, curls up its tail and begins to digest the food it has eaten. As soon as the Saturnites who are hunting this animal notice that it is resting, they must make certain that they overpower its tail as quickly as possible; they usually do so by severing the tail from the body with one stroke. After the tail has been severed, it is

advisable not to stay in the vicinity of the wounded animal, because the tail still has life for quite a while, and it throws itself back and forth in the most horrible contortions. The *bauor* itself enters into a rage and jumps and kicks with its hooves. The Saturnites are aware of this behavior, and that is why they immediately leave after they have severed the tail; they wait on their ships on the surface of the water until the animal collapses and remains motionless and its tail stretches out spasmodically.

12. Once these events have transpired, then our hunters approach the shore. They cut the tail into smaller pieces and, while they are doing so, the pieces still jump around for quite a while. Then they approach the animal itself and probe it by stabbing it in the area of the back to see if there is still some life left in it. Should the animal not react, they then remove the beautiful skin. But before they begin this task, they cut off the head and throw it into the water, because they believe that the animal could open up its terrible eye while they are skinning it, and all of them would be poisoned by it. That is of course impossible, because this eye does not contain any poison at all, and the powerful red light of the eye becomes completely extinct as soon as this animal has lost all its vital energy.

13. Once the animal has been skinned and the skin has been loaded onto the ships, they leave everything else to rot. The decomposition of the animal gets a helping hand from all the hungry guests that live along the shore, and for others it is an opportunity to revenge themselves on their well-known enemy.

14. For what purpose do the Saturnites use the skin, which is obtained with such great effort? The skin is treated with oil so that it remains soft. Once it has been properly treated in this manner, the skin is prepared and cut and then it is tailored into a half-length coat. A coat like that on the back of a man means more than an

entire kingdom. There is a saying on Saturn: It is the *bauor* which lends the prince or the most superior patriarch the proper respect and what he is supposed to represent.

15. Conquering a *bauor* is one of the greatest and most daring feats that the Saturnites can carry out. Whoever has caught such an animal and wears its coat shows to his fellow man the great courage he possesses. This testimony carries the greatest significance to the Saturnites because they feel that they cannot be properly served by a ruler or leader who lacks courage. Such a coat is proof to them that the person who obtained it under those circumstances was willing to sacrifice a great many things. This leads them to believe that the wearer of the coat, besides his great courage, is a very generous person because he did not spare any costs for the benefit of his brothers. And finally they attach to the owner of such a coat a great amount of intelligence, because he conquered this monster, which has to the Saturnites a considerably more terrible reputation than a dragon or lindworm on Earth.

16. If a person has conquered the *bauor*, then he should have the ability to deal with any other situation or undertaking with the proper intelligence. Therefore it makes the owner of such a coat unfailingly and immediately a grand patriarch, even if he is three or four times younger than any other minor patriarch. But he is only a grand patriarch as long as this coat lasts.

17. Since it ensures the honor of being a grand patriarch, nothing is cared for more than such a coat. That is why the *bauor* coat is worn by a grand patriarch under the most extraordinary circumstances. Since there are all kinds of little deceptions on this planet, this also applies to the life span of such a coat; the concept is that it lasts forever. Even if the coat is falling apart and can no longer be worn, it is replaced by a false one from the skins of other animals, and is paraded as a real *bauor* coat.

18. A grand patriarchate which has been acquired as the result of such a coat is inherited by all the children and the children of the children of the grand patriarch as long as the coat can be produced or its existence can be proven by showing it. Should someone else, however, come into the possession of a new coat by hunting the *bauor* and shows this coat to the people in his country, this ends the old patriarchate. The old patriarch however, still enjoys the position of an esteemed person. In a case like that, even the smallest piece of a *bauor* coat still represents a valid patent of nobility. This entitles the owner to *bauor*-coat privileges, but only as long as he can produce a piece of it, no matter how small. But once the moths have eaten up the last little piece, he then loses his position of arch-patriarch, and nothing more remains but an empty memory.
19. But at this point we shall not pursue the political conditions on Saturn any further, since we have not reached that point yet; and since we are not yet dealing with human beings, we shall return to the animals. Before we begin with the domesticated animals, we shall take a general overview of the entire wild animal kingdom.

29
Harmony Of The Celestial Bodies
Examples Of The Art Of Music
The Secrets Of The Teachings Of Sound And Creation
Correspondence Of The Wild Animals On Saturn And On Earth

1. You already know that we have only described those animals in detail which are extraordinary in respect to their species. From what has been disclosed thus far you have learned about the peculiar animals which are indigenous to this planet. When these animals

were described, you must have noticed that most were of such a kind that, as a whole, no animal on earth or on any other planet could be compared with them.

2. However, there exists a perpetual harmony in everything between one planet and another, without which two celestial bodies cannot exist, even if they are far removed from each other. To understand this properly, I must point out to you that harmony can only occur where one and the same cause brings forth the effect.

3. For example, if you were to tighten a string over a flat board and pull the string, it would at all times give you a particular high or low tone. Should you tighten the string more, then the sound would become more intense or, as you would express it, the tone will be higher. However, the less the string is tightened, the lower the tone. What is here the cause of this sonorous effect? You will probably find or specify no other cause than the board and the string which is tightened over it. As often as you renew the cause, that is how often you will have the same effect. The changes, in as far as the sound produced is low or high, make absolutely no difference, because a sound remains a sound, regardless of whether it is high or low. You might ask: What actually causes the sound — is it the string or the flat board? And I will tell you: It is neither the board by itself nor the string by itself, but both objects together. The flat board as a coherent whole has all conceivable forms for the formation of a tone at all times in readiness. The vibrating string above the board calls forth these forms in a manner distinguishable from one another. Therefore the flat board contains all conceivable forms of sound. The string which is stretched over it is there to awaken these forms of sound in order to transfer them into a perceptible appearance. For this to occur there must exist an undeniable harmony between the flat board and the string.

4. If you would consider the air as a means of forming sound, then it must be shown, when bringing forth any kind of effect, that it is impossible for more than two polar causes to be brought forth into a reciprocal connection. The means, however, cannot be considered a cause, but only a path by which the effect of the two polarities is brought forth to appear.

5. For example, observe the magnetic fluidum! Can the magnetic fluidum, when it is absorbed by an iron rod, only be found polarly? Or, rather, is it polarly freely effective within itself throughout the entire infinity? Therefore an iron rod is only the path upon which this fluidum can express itself tangibly to your senses. It would be an impossibility to consider the iron rod as that which brings forth the magnetic fluidum itself.

6. Or would you say that the air and the ether between a sun and a planet are the source of light? They are no more than the path upon which the light reaches a planet while emanating from a sun, provided that the planet is structured and is capable of absorbing the light that overcomes it.

7. Therefore, we shall not consider the air in this manner, when forming a tone as a means of being the cause of the sound, but only as a path upon which the tone-forms as they are developed between the string and flat board can be perceived by the ear.

8. Do not imagine the sound when you think about "tone," but only a form which, through a certain degree of vibration, is drawn out of some smooth and elastic surface. The sound is no more than a witness which, through regular vibration of some body that has the ability to vibrate, develops the forms of another underlying body. Even though you might believe that you are well-versed in the art of music, I say that there is hardly a subject with which you are less acquainted or more ill-informed about than the art of music. As

far as this subject is concerned, you do not know more than the worms which gnaw on the dead bark of a tree. Consequently, you compose a few different high and low tones and you take delight in them, just like the worms when they gnaw on the dead bark of a tree. But who among you has ever had the idea that the tone is one of the most wonderful forms?

9. If you sing a note or a tone, or if you generate sound with an instrument, you have nothing more to say about it other than: This tone is a C or an A, and it belongs either to this or that octave. And you may also have the ability to know which instrument produced that tone or sound. Admit it: You do not know much about a tone, except perhaps that you can assess the quality of a tone and can rate its proportion in comparison with your ear, but that is about the extent of your knowledge of the tone.

10. That you may have a fundamental understanding of how little knowledge you possess in the art of music, I will enlighten you a little in passing about the tone itself.

11. You know that many strings can be tightened over one board, and each string will, in accordance with its tension, produce a different high or low tone, and everything will occur on one and the same board. If all possible differences of sound can be brought forth on one and the same board, then there must be an infinite number of forms contained in that board, so that they can appear completely perceptible through each possible degree of tension of the string.

12. When you closely examine this particular board, what do you find? Nothing but an empty, flat board! And when you examine the strings that are attached to this board, what do you find? You find a uniform elastic string, either made of metal or the guts of an animal. And you have nothing but two flat uniformities, about which you cannot philosophize. Despite this, such multifariousness

lies within these two flat uniformities that all the composers who have ever lived, dating back to the times of David, have not even taken a one billionth[27] part of this in all their compositions; and these external tones are, in respect to the actual real tone, nothing but what the dead bark of a tree is in respect to its internal invisible spiritual life.

13. In accordance with this, what is a tone? Tone is nothing more than a self-expression of the many endless harmonious spiritual forms and how they are inherent in matter or how they project into this matter. Therefore, the board of a musical instrument that vibrates with that instrument is an infinite world filled with spiritual forms. For example, if you strike a C or an A, then an entire creation with an eternally innumerable number of beings of all kinds reports to your ear in a uniformly perceptible way through that rudimentary sound.

14. As a human being, you remain with what you perceive but you do not examine what is behind it. And even when, after several successive sounds, you are seized by some great ideas and these living spiritual forms literally grab you by the neck, you are blind and gnaw on the bark, without thinking that with every single tone, through the perceptible tone of a single word, all things that fill the entire infinity come forth. Now you should have a small idea of what a tone is and how different its great significance is from the monotonous sound that you call "tone."

---

[27] One milliard in Britain and Germany = one billion in France and the United States. – ED.

15. Since we previously proceeded from harmonious conditions and described how a continuous harmony exists between a flat board and a string, and how these effects at least outwardly originate out of this harmony, we can therefore bestow complete validity upon our first sentence, when it was said that between two celestial bodies there must exist a continuous harmony regardless of their distance from each other.

16. Why is that so? Think of the sun as the vibrating board and the planets as the strings that are tightened over it. Now when the planets or strings that float around the vibrating board of the sun are struck by the light that emanates from the sun, the planets take on all the underlying forms that exist on the sun after they have received this legacy by means of the light and make the forms appear outwardly.

17. If you now direct your attention upon the string of Saturn which is tightened over the same sun as the string of the earth which you inhabit, you should easily understand that the same cause which affects your earth and allows its forms to appear will therefore have the same effect on Saturn.

18. For example, compare a piano with seven octaves to one with only five; you cannot deny that the piano with seven octaves has much higher or much lower tones than the one with five octaves. However, when you start playing the scales on the piano with the seven octaves at the same spot where the highest or lowest tone begins on the piano with five octaves, you will find that the scale will have the same rising and falling sound as it does on the piano with the five octaves; of course the tones of the larger instrument will probably sound stronger, have more volume and be more developed than on the smaller instrument.

19. Actually, we already have all this information. I said in the

beginning that we would take a general overview of the entire wild animal kingdom on Saturn; before we describe the domesticated animals individually, I just wanted to tell you that we have already made such a general overview. The description of the productive power of the sun had to be mentioned before, so that what still has to be said does not appear as blather or as a compelling representation of things on this planet which make it appear as if the one who declares this has lost his imagination and therefore seeks refuge in what the earth has to offer as regards formal appearances. So that you would have to say: All the animals you find on earth will also be found on Saturn, though of course with a few variations; they are proportionately larger in size and stronger, and, as a consequence, the light of the sun is more refracted and the animals are more colorful.

20. Since such an anatomical analytical representation of the harmonious condition has preceded this, there should not be one person with a believing heart who would object when I say: Beginning with your largest original elephant down to the smallest mouse, Saturn has all of these animals as well on its surface, only they are proportionately larger and stronger and more varied in color between blue, green, red, white and black, while the colors of the animals on earth are rarely as completely developed, since the rays of the sun are still too intense and therefore they are not sufficiently separated when they fall to the ground. The coloration is at all times a consequence of the light. The flowers on earth are colored with all kinds of perfect colors, but these colors lack a certain living luster, whereas the flowers on Saturn become much more lively; not only does this coloration apply to the animals, but also to the humans on this planet.

21. This should suffice as a general overview of the four-footed

animals as well as the other animals of this planet. We will briefly examine only those few tame animals that the earthly piano with its five octaves does not contain.

30
The Tame Animals Of Saturn
The Most Beneficial Animal, The Cow, *Buka*

1. The first animal we shall observe is a large tame cow, which the Saturnites call *buka*. You might think: Why do we start with the cow and not with the bull? We are not dealing with a zoological account of the animal kingdom where, in accordance with the learned order of the so-called zoologists, the male must at all times precede the female. Instead, we are dealing with an account of the animals of the planet in the order of their usefulness and in accordance with their noteworthiness. Since the cow is much more useful and much more noteworthy, we shall deal with it first in its proper order; that means we shall deal with the cow before the bull.

2. What kind of animal is the *buka*? What does it look like, how tall is it and where is its home?

3. The *buka* or Saturn cow is a giant animal as far as its size is concerned; in spite of that, it is extremely tame. In accordance with its giant size it eats very little fodder, but drinks considerably more water.

4. With its plentiful supply of slightly yellowish and exceedingly good-tasting milk, this animal is the most beneficial among the creatures of this planet. Its milk is the main source of nourishment for the Saturnites. Now you would probably like to know how much

milk this cow produces in one Saturn day.[28] Since the Saturn day is not much different from a day on earth, it should not come as too much of a surprise to you when I say that this cow's regular milk production is frequently as much as 1,000 pails[29] in one day.

5. (NB! On earth such a cow would be quite welcome for some of your industrial tycoons, provided that it would not eat much more than a common cow on earth while it may drink as much water as it needs or wants. But since such very economical human beings would squander away too much if they were to possess such an animal, this animal shall therefore remain on Saturn — regardless of the fact that it would not be completely impossible to create a Saturn cow on earth!)

6. What does this cow look like? The shape of the animal closely resembles the earthly cow which you call aurochs.[30] However, its size cannot be compared; the Saturn cow is so large that a common cow on earth would not be any bigger than a fly on thc back of your earthly cows. After the *mud*, the male or bull is almost the largest animal on this planet. The cow, however, is considerably smaller than the bull. If such a cow were here on earth, and you were to stand on its back, you would have a much better view than if you were standing on top of *Plabutschberg*,[31] although the size of the cows on Saturn varies considerably.

---

[28] The diurnal rotation of Saturn is 10 hours and 14 minutes. – ED.

[29] One pail = 56.6 liters. – ED.

[30] The auroches or Ur, Urus (*Bos primigenius*) is now extinct. – ED.

[31] A mountain near the Austrian city of Graz, in the province of Styria. – ED.

7. The largest species of these cattle can be found most of all on the continent we know as Herrifa, which was mentioned right at the beginning when we began with the unveiling of Saturn. In this country such a cow frequently reaches a height of 2,400 feet, and it has twice that length from the head to the tail. Its body is supported by four proportionately strong legs, which, when compared to the rest of the body, are much shorter than the feet of a cow on earth when compared to its body. An extraordinarily large udder with eight proportionately long teats is located right between the hind legs. When the cow stands up, these teats are still 240 feet above the ground.

8. How can you milk such a large cow? You do not milk the *buka* like an earthly cow, because these cows give milk all by themselves. On account of their organism, the giving or the retention of their milk is controlled by the instinct-will of this animal. How do the Saturnites notice when this cow will give milk? They notice it by the puffed-up fullness of the udder and when the animal becomes quiet, which usually occurs after it has drunk a large quantity of water.

9. When the cow is at rest, the Saturnites make haste and place large containers with wide openings under the teats; these containers are from the pumpkins which we discussed in a previous chapter. The milk is carefully gathered in these containers, which the cow gives voluntarily. Once the *buka* has discharged all its milk, it lets that fact be known with a thunder-like grumbling.

10. When the milk gatherers hear this grumbling, they immediately take their containers and leave their station from underneath the abdomen of the cow, so that when the cow begins to move again no one gets trampled to death by the extremely giant and heavy foot of the *buka*. After a cow is several years old, you no longer have to worry about that, because an older cow will not move as long as

there is a human being under its belly. But this does not apply to the younger cows; they are by nature much livelier, and therefore more caution has to be exercised.

11. The Saturnites make quite a bit of butter, fat and cheese from this milk. They prefer to eat this type of food, especially butter and honey spread on cheese. On this planet, honey does not come from bees but from a species of flowers that have large calyxes and an exquisite fragrance; these calyxes are filled by more than half with honey.

12. Now we know how the Saturnites enjoy their milk. Therefore, what remains to be described is the color of the cow. The body is blue-gray and the area of the belly is completely white. Beginning at the body, the legs gradually turn dark blue; this applies to the front legs as well as to the hind legs. The tail of this animal is also darker in color than the body and is adorned at the end with an extremely strong vermilion bushy tuft. The neck is more slender than it is massive in proportion to the body. Beginning from the head to the front legs and on each side there is a long strong vermilion mane, the length of which is often up to 3,000 feet long. The head has no horns and it is small in proportion to the rest of the body. The bull, however, has two small upright horns which are bent backward like those of a chamois.

13. Something that accentuates the head of this animal are the ears, which have a length of 180 to 240 feet and a width a third of their length. The ears are dazzling white. The forehead of this animal is light bluish gray; in the area of the proportionately large eyes, this color is somewhat darker. The snout resembles that of a cow on earth; it is bare and dark gray. Everything else pretty well resembles the aurochs on earth.

14. Is this animal kept in a barn? It is too large to build a stall for

it. But it is kept in a living garden. On Saturn this is a large pasture fenced in with the wall tree. The cow cannot jump over this fence in spite of its size, because its legs are short and it cannot lift its feet any higher than 30 feet from the ground. This is, in other words, the stall for this cow. Naturally, such a fenced-in meadow is three times the size of Styria.

15. How many cows does an inhabitant of Saturn usually have? I tell you, anyone who has ten cows and two bulls is considered to be the richest.

16. That is pretty well everything which is noteworthy about this animal. Therefore, we shall now turn to another useful animal, the so-called domestic blue goat.

31
The Blue Goat
Bartering With Its Milk
The Goat Thanksgiving Festival
The Connection Of The Saturnites With The Spiritual World
The Wool Of The Blue Goat And Its Use

1. What kind of animal is the blue goat? It is an extremely useful and indispensable animal for those Saturnites who are of moderate means, particularly for those who live in the mountains, where the large cow cannot move about freely and where there would not be enough to eat, though mainly there is not enough water available there for the cow to satisfy its great thirst.

2. What does this animal look like? Does it look like any goat on earth? No, it does not. It somewhat resembles an elk, which on earth inhabits more the northern parts of the continents; this goat

is of course at least a hundred times larger than an elk on earth. It has between its two hind legs in accordance to its size a very large udder with six teats, which easily produce ten to twenty pails of milk.

3. The milk of the goat is not as sweet as that of the giant cow, but it is much more fragrant and more nourishing. Therefore, it happens quite often that the mountain dwellers take their good milk product to the people in the valleys and plains to exchange it for other things which are of absolute necessity to them. On this planet there is no trade other than barter. This barter serves a good purpose for the mountain dwellers. The goat cannot thrive in the valleys and plains, but it does at the heights, where it searches for food under the snow with shovel-like forward-bent horns. You should understand that on Saturn, as on earth, the highest mountain peaks are covered with ice and snow during the time of the ring shadow.

4. This animal is somewhat shy by nature, but when it is treated well by human beings it becomes tame and domesticated; it almost runs behind them like your loyal dogs do on earth. Therefore, when these mountain dwellers want to go somewhere they have to tie this animal to a long grass rope on a tree, otherwise the goat would follow them. The Saturnites, especially those that live in the mountains, have a festival every year which they offer to the Great Spirit in gratitude for this beneficial animal.

5. Quite a number of the most beautiful goats with full udders are taken to this festival. There they are taken to a particular spot where they form a circle and where the goats are milked, and this milk is gathered in the cleanest and most beautiful containers. As soon as this work is done, the animals are led to a rain tree pond which is always close by and, as part of the obligatory thanksgiving, they are watered with this exceedingly good-tasting pure water.

After that they are let loose so that they can graze on the succulent pastures which surround this rain tree pond. The people return to the place where the fresh milk is waiting for them in those beautiful containers.

6. Everybody takes his container and carries it to a temple which is specially built for this festival. The temple is planted with ray trees (*bruda*) or grown with mirror trees (*ubra*). The reason why I say "planted" or "grown" is because, on this planet, temples of divine service are not built by carpenters as on earth, but are grown out of the ground alive, and consist of the most beautiful trees. These trees are planted in a particular order almost like a cordon on earth in the gardens, which are regularly pruned. A temple like this, when fully grown, is a magnificent sight, especially around the time when the trees are in bloom; the sight is so magnificent that you on earth cannot imagine it. This temple is so large that it would take the people on earth a whole day to walk from the holy entrance to the opposite side where the exit is.

7. When the people have taken all their pumpkin containers with the milk into the temple, they first thank the Great Spirit for the gift of this beneficial domesticated animal and also for the milk they have taken from it. After this prayer, the eldest rises from their midst and asks all the devout who have gathered to lie on the floor face down.

8. The elder looks upwards and implores the Great Spirit that He may permit the elder to be overcome by a Spirit of Light and proclaim to him what would be pleasing to the Great Spirit, what they should do while in the sanctuary. And since the inhabitants of Saturn, but mostly the mountain dwellers, are almost in constant contact with the spirits of their heaven, it happens all the time that after such a request by the elder a shining light-spirit in human form comes to them and proclaims how they are to conduct themselves.

9. Once the spirit is finished with his proclamation, the people arise and the elder announces what he has been told. After the sermon the people give thanks again to the Great Spirit. Once that has been done with the greatest devotion, the people of both sexes fetch their milk containers and take them to the elder so that he may speak the blessing of the Great Spirit over them. Then they return with their containers, embrace each other, and everyone invites the other to share his milk and an assortment of other edible foods. After such an invitation, they begin to eat in the temple and have conversations that contain all kinds of information.

10. On this occasion, when almost all the food has been eaten, throughout the day they give thanks again to the Great Spirit, which the Saturnites seek to heighten through the songs of the birds whenever they are available. This is not done by the main singer but by the singers of the second kind, as previously mentioned.

11. After this thanksgiving everybody leaves the temple, but, mind you, never by the holy front exit, but always by the rear exit, which is intended for the people; whereas the holy exit is only for the elder and for the Spirits of Light. As soon as the people are outside the temple, they call their goats which are grazing leisurely around the rain tree pond; the goats obey the call and follow their masters home.

12. This is the simplest of celebrations for the Saturnites. Whatever concerns their main celebrations and the main divine service shall be described when we discuss the Saturnites in particular.

13. When the inhabitants come home with their goats, they milk them again and then let them roam free once more, for the Saturnites build no barn for these goats. Also, nobody makes a goat his permanent possession; if a goat comes to the dwelling of a human being with a full udder, they milk it and let it go. Also, nobody has

to care for a goat, feeding it, etc., as it does not require anyone to tend it. These animals take care of themselves; they are so tame, good-natured and attached that they always come at the right time to the dwellings of the people. They also do not need a guard or attendant, because on Saturn, especially in the mountains, there are no carnivorous animals.

14. If you remember, those animals which are wild and antagonistic usually live in areas which are far from the continents inhabited by human beings, and they are separated by water. In the countries on the continents, these wild animals only inhabit areas where humans never set foot. If that should happen, it will happen only because of curiosity, pertness, and sometimes also greed. But in the mountains, as you know, wild animals other than the "pointed foot" (with which we are already acquainted) very rarely live.

15. From what has been said, you can gather how easy it is for the Saturnites to keep this animal and how beneficial it is for them. We have now learned everything important about this animal.

16. I do not think it is necessary to reiterate why this animal is called the blue goat. Just as the name indicates, its color is blue. In addition, this animal has extremely fine wool in plentiful amounts, like sheep on earth. The Saturnites, especially those who live in the mountains, make all kinds of warm clothing out of it for the colder shadow season. They clean the wool beforehand, then they spin it into beautiful, uniform thick threads and they weave all kinds of differently patterned cloth with very peculiar kinds of proficient tools.

17. What happens when the blue goat dies? They skin it and the meat is placed into a pit, because the people on Saturn hardly eat any meat.

18. That is everything there is to know about this animal. In the next chapter we shall describe another noteworthy domesticated animal.

32
The Saturn House Servant, *Fur* – A Species Of Tame Ape Concerning The Saturn Dog, Horse, Sheep And Other Animal Species

1. As far as this animal called the *fur* is concerned, its name is derived from the purpose that it serves for the Saturnites, whereas the animal which we discussed previously derived its name from its color.

2. What does this animal do? It renders the same services to its master as a faithful and diligent house servant on earth. This animal carries out almost all the work which you on earth would consider to be heavy or hard work with the greatest accuracy. This animal tills the fields, carries water to the house, gathers wood and carries it to the house for the human beings to use, cleans the fields and chases away harmful animals, and at night guards the entire household and carries out many more such chores.

3. Therefore the people on Saturn have given this animal the name *fur* which is in accordance with its usefulness; the meaning in your language is "faithful house servant."

4. Now we already know what this creature does and what its name is. What does this useful domesticated animal resemble? That of course is an entirely different question. Is there on earth a similar creature which it resembles in form? The answer is yes, on earth there are similar creatures, and they exist in large numbers in all

kinds of gradations. Only on earth these animals are without exception wild, whereas on Saturn the reverse is the case; there most animals of this species are tame and they learn so easily that in a very short time they can be trained to do all kinds of human labor and can be employed in this manner. What is the name of this animal on earth, which on account of its natural ability and physical structure could also be employed for most of these chores if one only knew how to domesticate it and teach it the various human tasks? On earth these animals are called apes; and particularly those which you know as the orangutan.

5. As aforementioned, these creatures are the most useful members of the animal kingdom on Saturn, because they can be employed for all kinds of work. Does it cost much to maintain such an animal? They do not cost much at all. They are the most inexpensive servants for the Saturnites, because they demand nothing but to be treated properly and to eat the occasional fruit from the hand of a human being. That is everything that these workers demand from their masters.

6. Only when they are abused or treated cruelly do they usually revenge themselves by becoming unfaithful and leaving the house forever. If they are retained by force, then the owner is in for a big fight wherein the animal always becomes the winner by running away. When people on Saturn offend the *fur* so that the animal wants to leave, they indicate to the animal by holding some fruit in their hands and offering it that they regret their error and want to make up; usually the offended animal turns around and, just as before, becomes again a faithful servant to its master.

7. What kind of food does this useful animal eat? Its usual nourishment consists of all kinds of lower-growing fruit on trees and shrubs, which is rarely eaten by human beings and not at all by

those who are well-to-do. From this you can gather how little such servants cost their masters. And if you add to this the fact that this kind of fruit is in great abundance and not fit for human consumption and that I care for the indestructible clothing of these workers, then you must clearly understand how extremely inexpensive these laborers are for the Saturnites.

8. Since we have mentioned quite a bit about the usefulness of these animals, we will now turn to the shape. I mentioned before that these animals closely resemble the ape on earth. However, on this large planet everything is much more developed and perfected in accordance with its shape; therefore these faithful house servants are much more developed and perfected that an orangutan on earth.

9. As far as the skin of the Saturn orangutan is concerned, it is covered thickly with hair except for the palms of the hands and a small part of the face. Also, the hands and feet are, as a rule, significantly slimmer and therefore less fleshy than those of a Saturnian human. The human beings on Saturn are all very well built and perfect, without any hair on their bodies, except for the head and genital parts. The color of the skin of the Saturnites is mostly snow white and, in the plains, the color touches on a reddish brown, whereas the color of the hair of this animal is either light-blue or sometimes gray. The parts which are without hair are at all times a pale red.

10. How tall is such an animal? It is close to the height of a Saturn woman; but nobody has ever seen one reaching the height of a Saturn man.

11. What is the main habitat of this animal? If you take into consideration that on this planet all the continents are in one and the same climatic zone, it should be easily understood when I say that this animal's habitat is on almost every continent, and mainly

with mountain dwellers, and that this applies to the entire planet with very little variance in its form and color.

12. This species of ape is not the only one that lives on this planet. There is an almost immeasurable variety of species of this particular animal, but all of them are wild. On some continents these animals are so numerous that they move in large herds to the mountains and, while there, they do not spare the noble fruit intended for the mountain dwellers.

13. When this occurs the faithful house servants perform an invaluable service, because as soon as they catch sight of these herds anywhere close to the fruit trees of the Saturnites, they immediately drop whatever they might be doing and run incensed towards these uninvited parasites. Woe to those intruders who are caught! They will not escape; instead they are torn to pieces immediately.

14. Because all the smaller species of this group of animals knows instinctively that the reception by these larger members of their species is anything but friendly, these confrontations happen very rarely, only when they are forced by vital necessity. Such a venture is always undertaken at night and never during the day, and only when the area is under the shadow of the ring (which to the people on Saturn means as much as winter on earth).

15. Now we know everything noteworthy about this animal, and for this reason we shall describe another domesticated animal which is held in great esteem by the Saturnites of the plains and valleys.

16. What kind of animal is this? In many aspects it resembles your domesticated dog. As far as its usefulness is concerned, on account of its strength and agility it is used for everything for which you use horses, except for riding, because the Saturnites do not ride animals. A Saturnite considers it below his human dignity to mount the ignoble body of an animal with his noble body; and nothing will carry him faster to his destination than his own two feet.

17. This animal also has many various gradations or species, which on the different continents exist in a variety of species, and, with the exception of a few smaller species, almost all are used for the same purpose.

18. These are the least beautiful animals; their color is similar to the color of other animals on Saturn, but they are not as clean and not as lively. Therefore, between dogs on earth and on Saturn there is almost no other difference other than size. The largest of this species on Saturn is 500 times the size of your largest dog on earth. But as far as the rest of these dogs is concerned, they are used besides their other activities as guard dogs. They do not bark; their voice is more like a thunder-like growling. The growling is of course much louder when it comes to the larger species, whereas it is not as loud when it comes to the smaller species.

19. That is about everything noteworthy that can be said about this animal. If you would ask the question: Is there one animal on Saturn that resembles our noble horse on earth? My answer to that question would be yes, there is a kind of horse on Saturn, but it is never tamed, because on Saturn it belongs to the wild animals.

20. Are there any sheep on Saturn? Yes, there are, but they are not tame either, but are considered as wild. And they are often hunted for their beautiful, soft hides.

21. There are quite a few of these species on Saturn in the form which are tame on earth, whereas on Saturn these animals are not domesticated but live wild.

22. The entire animal kingdom on Saturn has been described in the shortest possible manner and if you exercise your powers of imagination just a little, it should be easy, as a result of this, to achieve a very pictorial representation that you can almost imagine any animal described here and how it exists in its natural state on

this planet. The great multifariousness shall give new proof of how wonderful and abundant My infinite works are. And since these works exist already on one planet in such multifariousness and beauty, how many more wonderful and grandiose works does a sun contain on its vast soil, and how many unpronounceable, wonderfully grandiose and multifarious works does a spiritual world contain, when a material, natural world by comparison hardly represents much more than the external dead bark of a tree?

23. Information like this and many other things in comparison I shall disclose to you when I describe the people of this planet, as for today we shall conclude this chapter.

33
The Saturnian Human Beings
Their Origin, Population Density And Housing Conditions

1. Much has already been said about the humans on Saturn. Regardless of that, much more remains to be said about the masters of this celestial body, in order for you to learn in what kind of order this planet exists and whose Spirit's child it is.

2. Since a good order is at all times and everywhere the basis of all wisdom, we shall therefore observe here a human being in a good order — first of all, in his externally-shaped being and then his spiritual existence and everything that enters into the region of the spiritual, such as his country's constitution, his trade and his divine service. Now we shall begin with the physical shape of the Saturnian human being.

3. Was there only one human couple on Saturn, or were there at the same time at various places several human couples created? All

human beings that now live on Saturn originate from one human couple; this applies to all planets just as it applies to earth. The history of the Saturnites is more than one million earth years older than the history of human beings on earth.

4. A Saturnite throughout his entire life has rarely more than four descendants with his wife; it should be understandable when I say that this planet, in proportion to earth, is by many times less populated than earth. On the large continental countries, of which several are larger in area than Asia, Africa and Europe, human beings very seldom live in a neighborly way next to each other. This situation that exists in your villages on earth belongs to one of the greatest rarities on Saturn.

5. If a human being from earth were somehow able to travel to one of these continents, the dwellings of the people on Saturn are so far apart that it would take him at least ten to twelve days to travel from one to another. In the mountains these distances from one dwelling to another or from one neighbor to another are the rule rather than the exception, whereas in the lower-lying regions, which are next to the lakes and rivers, the dwellings of the Saturnites are closer together.

6. Wherever you find such an abode, there are not only a few human beings, but an entire family which usually numbers anywhere from 1,000 to 5,000 people.

7. How large are their dwellings that they can accommodate so many people? It should be mentioned that the Saturnites prefer to have plenty of space. As far as their accommodation is concerned, this was mentioned right at the beginning with the description of the sun tree (*gliuba*). It was also mentioned that this tree is their house, one which they prefer to all others. Therefore I do not have to describe the peculiar shape of this tree again, other than to

remind you that it is an extraordinary tree with many trunks, and that on its wide branches, which almost always run in a horizontal manner, the people build their dwellings.

8. Such a tree is as significant on Saturn as an important city on earth. The patriarch of such a large family assigns the individual branches and side trunks to the individual family and it then becomes their own property, just like a house in your cities. The difference here, however, is that the people who live in this tree city are all blood relatives, while in your cities anyone who has the financial means can purchase a house right next to you.

9. You might ask yourself: How can these people sleep on these branches? If they turn around during their sleep will they not fall out of the trees? The possibility of this happening would be much greater if you were simply to turn around in your bed. Because these horizontal outward-bound branches are so thick and wide that on a single branch there is enough room for every house in your capital city,[32] and there would still be enough room to drive and ride around.

10. Besides that, there are a large number of smaller branches growing in a horizontal parallel direction from the wide edges of each of these large branches; they are the actual carriers of the fruit of this tree. Closer to the trunk these smaller branches are of significant strength; they are so strong that if a human being would walk, stand or lie beyond the edge of the branch he would not fall off the tree. Let us assume that despite these smaller branches it happens that someone walks carelessly beyond them and falls to the ground, or from a higher branch to a lower branch; this person would not sustain any injury from such a fall. The reason for this is already

[32] This capital city is Graz in Austria with a population of over 200,000. – ED.

known to you, and has been mentioned in a previous chapter: On Saturn, through the existing attraction between the actual planet and the ring, the specific weight of any given body, including the body of a human being, is significantly reduced. Therefore, nobody on Saturn will be able to fall with the same severity as someone on earth.

11. The above explanation is the reason for this occurrence and should completely satisfy the inquisitive mind in this respect. As long as this planet has been populated by human beings, no one has ever broken a leg or an arm during his lifetime, and nobody has injured his head or sustained a hole in his head through a fall, which of course happens quite often on earth.

12. The question now arises: Is this tree the only dwelling the Saturnites call home? The answer is no. Besides these main tree dwellings, they have various other dwellings, which are occupied during the cool shadow time.

13. These houses are built in various ways. Some are built from the strong branches of the pyramid tree, while others are grown from the species of the more slender trees. The houses that are grown, or living house, are superior to those which are built. However, they always build houses between those which are grown, because they serve as storage for their food.

14. They also only make a fire in those houses which they build, where they cook, boil and roast their various food, though not in the same clever and artificial manner as you do on earth, but more as you would bake an apple or cook pears, vegetables, and some of your good-tasting beets and roots. This is what the entire art of cooking consists of on Saturn. In these houses they also store their milk products as well as their precious berry juice containers.

15. Nobody eats or sleeps in the houses that are built; that is

always done in the houses that are grown, because it is unbearable for the Saturnites to remain for a longer period in the vicinity of a dead thing. That applies to a house which is built from trees that have been cut down as well as to dead human beings or animals. That is why houses built by the Saturnites are used exclusively as workshops and storage rooms.

16. You would probably like to know how large such a man-made house is and also what it looks like. I can grant this wish. These houses are built in the form of a wreath, similar to the houses in the Austrian countryside, except they do not have any roofs; they are completely open towards the sky. For the Saturnites, the greatest atrocity is a house separated from heaven by a roof. They say that everything that comes from above unto the ground is a blessing from heaven, and therefore a blessing for the soil. And since they also live on the soil, why should they separate themselves and hide from the blessings of heaven? They are of the opinion that it is also a much greater benefit to them, since they are alive and therefore in much greater need of heaven's blessing than the soil of their earth, which they believe to be dead.

17. Now we know how their houses are built. Now all we need to know is their form and size. Usually these houses are built in the shape of a star, like a compass dial; however, these houses have sometimes eight, sixteen or even thirty-two pointed tips, each of which forms a kind of separate section for food and drink. In the center of this round wide room is a round hearth where food is cooked. It is understood that the hearth is in proportion to the size of the Saturnites.

18. In such a star-house the distance from one tip to the opposite tip, in accordance with your earthly measurements, would take one hour if you walked it in good stride. As far as the height of this

house is concerned, every person on Saturn that is as tall as a man can easily look over the walls of this house.

19. Are these houses decorated in any way? No, other than that the people adorn them with the foliage of the trees they cut down to build them with. That is everything in regards to these houses.

20. Since we have learned that the houses they build are the actual farm, kitchen and storage buildings of the people on Saturn, we shall now take a closer look at their living domiciles.

21. What is the form of these live domiciles? Their external form is completely round, with one entrance from the morning (east). Only two different kinds of species of trees are used to build or, rather, grow these houses. The more beautiful and more magnificent houses are grown with the well-known mirror trees (*ubra*) planted closely together; the houses that are less beautiful and less magnificent are grown with the wall tree (*brak*), not the common kind but the cultivated wall tree.

22. The inside floor of these houses is made completely level. Grass seeds are sown on this floor, from which a very short but extremely thick grass grows, like a thick carpet. This grass has the appearance of velvet and is so resilient that, when someone walks on it, it immediately springs back to its original state.

23. In the center of the house is a large, round and proportionately high hearth, which is covered on all sides with the same grass. So that you have an idea of the size, circumference and height of this hearth, I can tell you that it has a diameter of roughly 540 feet and a height that reaches above the knees, even halfway up to the thighs of a Saturnian man. However, when its height is compared to the height of a woman, the hearth is about half her size.[33]

---

[33] A woman is 80 to 90 feet tall and a man 95 to 135 feet. – ED.

24. What purpose does this hearth serve? It serves the Saturnites in the same manner as do tables on earth, namely for food and drink.

25. Not far from this hearth-table, at a distance of approximately 200 feet, there is a completely round pyramid, obtuse at the top, which at its base has approximately three times the diameter of the hearth. The space at the top of the pyramid is only large enough for a man to stand on conveniently. This pyramid has the height of a man and is covered with the same grass as the floor and hearth, and it serves as the family pulpit for the patriarch. Every day before sunset the eldest member of the family stands on this platform, whereupon the whole family gathers around in order to hear from him the will of the Great Spirit pertaining to the night and the entire following day.

26. What other kinds of furnishings or facilities does such a house contain? Completely in the background, the entrance directly opposite the sunrise, there is another circular wall covered with the same grass. But it is not as high by far as the center pulpit; instead, it is considerably larger in circumference and with many moderate curvatures. What is the purpose of this third circular wall? This is the common bed or place of rest for our tall Saturnian humans.

27. Before they go to sleep, they place soft pillows into the curvature close to the periphery of this circular wall and lean against one of these curvatures, each person having one such curvature. The men take their position, which faces the rising sun, and the women face the setting sun. All are resting with their bodies in a position that is 30° in earth measurements in relationship to the floor. That is how they fall asleep; they remain in this position until shortly before sunrise. In spite of the ring shadow they notice the sunrise, because the ring never covers the sun so completely that nothing can be seen

of the sun. And should it happen at times that the ring is wider that it covers the sun completely, the duration of such cover never lasts longer than half a day; after that a small edge of the sun becomes visible again.

28. These are all the furnishings which such a house contains which is occupied by the Saturnites during the time of the shadow. What is the size of such a house in regards to its circumference? The diameter is longer than the well-known star-like storage houses by more than half.

29. Does the entire population of the large dwelling tree live in such a house? No, only one family, that is one father and one mother with their children and children's children, just as they live together on one branch of a tree during the light-time. There are as many man-made dwellings as there are branches on which the Saturnites live.

30. In such common family dwellings, the area around such a tree covers mostly more square miles than Styria (632,586 square miles). These dwellings are, as you know, at such a distance from each other that it would take many days' travel to reach another one. Around these dwellings are the grounds and grazing lands of the domesticated animals which you already know. These animals must have enough ground so that enough can be harvested for the maintenance of the life of the people on Saturn as well as the preservation of the animals. In addition to this, there are, at border regions of the common family properties, the large, far-expanding forests of the funnel trees, and on the large plains, especially on the northern side of large lakes, there are forests of pyramid trees covering areas of 2,000 to 3,000 square geographic miles, and there are also other large forests of shrubs. When you take all this into consideration as well, you will not be surprised that the distance between two common family dwellings is so far.

31. Now we know how the people on Saturn dwell, especially those in the mountain regions. But we do not know anything about the rules or laws by which they govern their households. However, we examined those Saturnites and their dwellings who live in the mountainous areas. But before we turn to the people who live on the plains, we shall first examine, in the chapter that follows, the household constitution of the people of the highlands.

34

The Household Constitution Of The Saturnites
The Planting Of A Temple
The Most Important Law: Emigration
Love Thy Neighbor Or Charity

1. Who is the head or principal of such a large family or commune in the highlands?

2. If it happens that the progenitor is still alive, he remains both the head and high priest of the family for as long as he lives. When he dies and there are two or more sons, the oldest son is chosen to be the head in domestic as well as in spiritual matters. When the eldest son dies, and one or more brothers are still alive, then the oldest surviving brother always becomes the head. When the last of the brothers has died, then the firstborn son of the oldest of these brothers becomes the head; in other words the brother who took over as head of the family from the progenitor. Under these conditions the oldest in the family will always be the head of the family.

3. When a family is not as numerous, then the transfer of the duties as head of the family from one member to another sometimes continues up to the seventh generation, in some instances even to the tenth generation. What happens when the family becomes too

numerous, when it reaches the fifth generation? At this point a separation takes place which happens in the following manner: the eldest brother takes over the family responsibilities, whereas the two younger brothers, after they have been given a dowry by the brother that remains as head of the family, take their family members and leave with their belongings in search of another uninhabited dwelling tree (*gliuba*). After having found such a tree they say their thanksgiving prayers and, under the leadership of the oldest, beseech the Great Spirit to bless and maintain the live domicile as well as all their other dwellings.

4. Once this prayer has been completed, the oldest takes several steps to the side and prays by himself. He is asking the Great Spirit to provide him with a Spirit of Light, the same as the Great Spirit did for his father and forefathers, and that this Spirit may at all times proclaim to him the will of the Great Spirit. During this prayer, all family members lie on the ground with their faces down. The oldest does not cease with his invocation until the Great Spirit has sent him the desired Spirit of Light.

5. When the Spirit of Light has appeared, the elder asks the Spirit to bless the new, uninhabited tree in the name of the Great Spirit, and then to take him first into the tree to show him the place which he is to occupy as the head of the family. Once that has been done, the oldest gives thanks to the Great Spirit in the presence of the Spirit of Light for this great blessing. Then the oldest is led down by the Spirit of Light to the spot where the Spirit of Light first appeared to him. Here the Spirit of Light leaves the oldest, but not until he has strengthened his will.

6. Strengthened in this manner in his Innermost, the oldest returns to his family, which is still lying on the ground face down, and speaks to them in a strong voice, whereupon they rise, praising

the Great Spirit for bestowing such a great blessing upon them and giving them their own enlightened patriarch.

7. Once that has been accomplished, the oldest assigns the branches to the fathers of the families, and they take possession with gratitude. The branches are cleaned and made ready for occupancy.

8. On such an occasion, which does not occur too often on Saturn, everybody is very happy and cheerful. They have a livable tree, but they do not yet have their live houses to live in and the storage buildings. That is why, on the first day, not much is achieved and everything is properly taken into consideration, planned and discussed, but always under the leadership of the oldest, because nobody will do anything without his consent.

9. At dawn the next morning, they immediately begin measuring the areas for the houses which are yet to be grown. Then the oldest blesses the lots, and the seeds of those trees which are appropriate for growing such live houses are placed into the ground in the proper order.

10. Once this work is done, which rarely takes more than one day, they go into the forest the next day to cut the proper wood for building the storage buildings; the domesticated animals also help with this work. *Horud*, the Saturnian lion, which is half-tame and half-wild, cuts down the biggest tree branches from the pyramid tree with its strong beak and the well-known house servant *fur* picks up these branches and carries them to the place indicated by the people.

11. As soon as all the required wood is at the building site, it is properly trimmed and the storage buildings are built.

12. When that is done, the animal stables and gardens are planted. Also, a fairly wide dam is built around a rain tree so that water can be gathered and a pond can begin to emerge.

13. Should there be any mountain springs, they build simple water conduits through which water is conducted to the main dwelling. These conduits are built with the aforementioned stick snails. Should these snail shells not be available, the fruits of the funnel tree are used.

14. Once that has been done, other properties are measured and their use determined. And when this allotment has been completed, should there be forests of funnel trees too close to these properties or infringing upon these properties, the trees are cut down and the forest is cleared to the extent that they have the right size for the property. When this work is commenced the aforementioned animals are given their tasks. The wood is piled up to dry at the end of the properties and later is used as firewood.

15. When this work is complete, the properties are blessed by the patriarch and all kinds of fruits are sown. This seeding usually takes place every ten years on this planet. However, one seeding fully suffices for all future harvests where the soil is richer, because the roots of Saturn plants do not die off so easily, but remain alive in the soil, like the roots of so many shrubs and bulbous plants on earth.

16. Once this work is completed, it is the last work in regards to preparing the soil for the harvest; then the family gets together to give a prayer of thanksgiving to the Great Spirit and at the end of this prayer they beseech Him to bestow upon all these fruits and upon all their labor a prosperity that has His blessing.

17. Once they have completed these prayers of thanksgiving and supplication, they begin with the most important work, namely the planting of a temple, wherein they offer the great God alone a sacrifice that is pleasing unto Him. This is performed by the patriarch and two of the elders and no one else is allowed to put his hand on this holy work.

18. Where do they begin with this work? On this occasion the patriarch proceeds to the location where the messenger from the beyond first appeared, and he beseeches the Great Spirit to most graciously show him, through the Spirit of Light, where He, the Great Spirit, would like to have the temple erected. When the patriarch has implored the Great Spirit long enough, and no one has appeared, then they build the temple where the Spirit appeared the first time. Should the Spirit appear, he leads the patriarch to the spot where the temple is to be built, or the patriarch sees the Spirit already on the spot where the temple is to be built. The patriarch goes to the spot where the Spirit awaits to show him the outline of the temple.

19. On the spot where the Spirit is waiting, a mark is placed on the ground; that is where the elevation is built from whence the patriarch will teach his family. At the same time he is shown a spot by this heavenly messenger in the direction towards the rear exit, upon which the patriarch, after appealing properly to the Great Spirit, will learn the will of the Great Spirit from the heavenly messenger.

20. Once all this is done, the Spirit will become invisible. The patriarch gives a sign, so to speak, from man to man through a number of messengers who are waiting within a certain distance from each other right to the dwelling tree, with the message that he has permission from the Great Spirit to build a temple on that spot. He asks them to thank the Great Spirit with him for such a blessing and to beseech Him that seeds to grow the temple may soon prosper, and that He may find all of them worthy to let them know His Holy Will.

21. This prayer is very solemn, and as soon as it is completed the patriarch calls for two or three of the oldest and distributes seeds

among them that have been breathed upon by the Spirit; these are planted in the soil to grow the temple. The elders immediately begin their work and plant the seeds of the most magnificent and beautiful species of trees with the greatest devotion and confidence.

22. The two or three eldest plant the front part of the temple which is intended for the people or family members, whereas the patriarch plants the sanctum of the temple, which consists mostly of ray trees *(bruda),* while the other part of the temple is usually made up of mirror trees *(ubra).*

23. Besides the elliptical egg shape of the temple, a circle of the noblest kind of wall tree *(brak)* is planted at the proper distance instead of a ring wall. This noble kind of tree can be distinguished from the common kind by its bark. The bark of the common kind looks like gold, whereas the bark of the noble kind looks like many rainbows one on top of the other. The colors have a very lively metallic luster. The leaves on the upper edge have almost the form of aloe leaves, only their size is in proportion to everything else on this planet. Their color is a dazzling white. The blossom is just like the common wall tree, only more delicate with an extremely lovely fragrance.

24. When these temple builders have completed their work, they again thank the Great Spirit for the energy and the insight that enabled them to build this temple, and they ask Him to help the seeds so that the temple may grow perfectly from the planted seeds.

25. After these prayers of thanksgiving and supplication, they leave the building site with great awe, and walk backwards more than halfway from the wall tree. They bow and then depart for home.

26. When they reach their dwelling, they ask the others to get up from the ground and climb up to their allotted branches of the tree

they dwell in, upon which they eat and drink after the eldest in the family has blessed their food. Because, while the temple is being seeded, which for the Saturnites is one of most spiritual rituals, neither food nor drink is consumed.

27. After the meal, which is under those circumstances as well as on a specific spiritual day celebrated in the evening, the elder admonishes all family members that they should sincerely unite their will with the will of the Great Spirit and that they should have no other will now other than that the seeds of the temple should soon prosper in a wonderful manner.

28. Upon such admonition, every Saturnian inhabitant present hones his will, invigorating the soil where the temple is seeded. And you can believe this: the circumstances surrounding such an occasion are absolutely miraculous; the seeded temple very often stands there complete in all its great splendor on the following morning, which to you might seem inexplicable.

29. When the family sees this temple, their devout jubilation and praise finds no end. This jubilation, praise and glorification sometimes lasts for several days and nights.

30. Why are these people so happy when their temple has grown so well and so fast? There is more than one reason for their happiness. The main reason is that they are convinced now that the Great Spirit is with them in their new domicile as He was before in their old one. A second reason is that, through this, they once again have a place where they can and are permitted to approach the Great Spirit through their patriarch. And a third reason is that this family, which is now separated from the main family, through this temple is obviously pleasing to the Great Spirit.

31. There is still another reason, but one which is closely related to the third reason, which is that this entire community can now

draw the conclusion that the new property they occupy is rightfully theirs and is therefore a place of permanence. Because when a temple does not grow as quickly, then from their point of view the property is not rightfully theirs, and therefore if someone should come and say that this property had been in their possession before, they must at all times be prepared to relinquish it immediately without any opposition and look for another property elsewhere.

32. Once there is a temple, any former owner loses all rights to this property. He would not, at the sight of such a new temple, make even the slightest demand for the restitution of his property. On Saturn the greatest law in existence is as follows:

33. Whatever the Great Spirit has given to someone belongs completely to those who have received it from Him. And no being of heaven, no Spirit of Light and no worldly creature has the right to contend this highest of all gifts. Whosoever would do that must be cast out to those regions of the planet where nothing else awaits him but eternal cold, eternal ice and eternal snow.

34. Every person on Saturn knows this law. And there is hardly any law which is more respected than this one. For this reason there are never ever any ownership disputes, especially among the mountain dwellers, because among themselves they always observe the best of order.

35. In regards to the relationships between neighbors, there are no real borders; instead, wherever someone settles, that is as much as his property, because every human being on Saturn carries as much as the testimony of the Great Spirit within himself, and this suffices him to travel anywhere on this entire celestial body.

36. Should it happen that a continent occasionally becomes overpopulated, several families will join together and leave their homeland and, on the ships we discussed in another chapter, move

to another land on Saturn. Once they have reached this land, they look for suitable accommodation. When they have found such a place, they have the right to remain there for an entire year and partake of what the soil has to offer.

37. Should the seeding of the temple prosper in the manner previously described, or should the temple grow gradually in the natural manner, and if the emigrated family can show the building site of the temple to the emissary of that country's natives who has been dispatched to them for that purpose, and if they can prove to him that all the temple trees are all growing well, then under those circumstances they are the owners of the property which they occupy. And the previous owner no longer has any rights to the property of which the newcomers have taken possession, except the right to friendship.

38. A human being on Saturn never denies this right of friendship to another; they immediately establish a relationship. Whereupon the patriarch of the newcomers says to the emissary of the native people:

39. "Brother in the Great Spirit, observe how your eyes examine me, but do not find anything which would deny them access to my person; by the same token, your heart shall not find anything in my heart that will deny you anything of which you are in need; that is why you are a brother to me in the Great Spirit."

40. After these words have been spoken, they embrace each other. This embrace is a permanent document of complete community property of two of such families. Whereupon the emissary invites the entire newly arrived family to make use of his dwelling, until the new property has grown in all its aspects. Soon thereafter, the newcomers follow the emissary, who is usually an elder himself, to his house.

41. Such an occasion is at all times one of great celebration. Because for the Saturnites there is almost nothing greater or more exalting than finding a brother in the Great Spirit on another continent or in the country to which they have emigrated. This is how the Saturnites address each other.
42. Should it happen that the newcomer, upon his arrival in the dwelling of the other elder, finds that his host lives under very poor circumstances, he offers his host his services to cultivate and extend the grounds; these services are always graciously accepted by the host and, in return, the host offers his services to his neighbor.
43. If, however, the newly arrived patriarch says to his new neighbor: "Brother in the Great Spirit! I am now convinced that you are very needy; therefore I have decided to return your former property to you and I shall look elsewhere for a place to live," the answer to this offer by the native patriarch will then be as follows: "Brother in the Great Spirit! I would rather cease to live and wish not to exist before I would see you leave the place which you have taken with the permission of the Great Spirit. You know very well that it is not the soil but only the Great Spirit which gives us all the means to live. Therefore, the ground we live on is large enough to completely nourish at least ten or more families like ours."
44. When the newcomers decide to stay there is another celebration, and the original inhabitant does everything in his power to have his new brother for all times close to his brother-heart.
45. This is all for today. In the next chapter we shall continue to deal with the constitution of the mountain dwellers.

35

The Main Law Of Life: The Will Of God
Treatment For Offenders

1. So far we have learned that on this planet, namely among the mountain dwellers, there are no borders as far as personal properties are concerned, and that the expression on the face of a human being is sufficient testimony for the Saturnites that the Great Spirit has given him the indisputable right to take possession of such properties anywhere on the planet which suits his needs. We shall now return to our family which was separated from the main family under the leadership of a new patriarch.

2. We have seen how the new temple was built and we have heard about the possibility of how such blessed seeds grow with miraculous speed from the soil of this planet, and we have also seen how the Saturnites build or grow other buildings. This completes the description of the new development of their property. We shall now describe what you on earth would call a political constitution.

3. What does a political constitution consist of when it comes to such a family? On Saturn a political constitution is very short and contains only a few words; the foundation of this constitution consists of only one main law, namely that not one member of any family community is permitted to do anything unless the patriarch communicates to him the will of the Great Spirit; neither will any community member do anything unless he is told. Once he has been told the will of the Great Spirit by the patriarch, he cannot begin the work until he has given thanks to the Great Spirit for proclaiming His will; after giving thanks he will also beseech the Great Spirit that the work he is undertaking will succeed in a rightful and proper manner.

4. That is the main foundation of the entire political constitution of the Saturnites. Every human being acts in accordance with this principle and does not worry about anything else; the Saturnites care only how they can properly thank the Great Spirit after they have completed their work.
5. You can believe that this short sentence contains everything imaginable. Whosoever acts in accordance with My will acts properly at all times.
6. Therefore there are no additional interpretations about this brief political law, which any child can commit to memory after it has been repeated three times. And this short law has no penal or criminal codex at its side as a political taskmaster, except for the expression: "I act in accordance with the cognizable will of the Great Spirit!" This expression is for every human being on Saturn the most powerful documented evidence of lawful conduct, which cannot be influenced by anything else.
7. Should it, however, occur that at times when someone from the depths[34] comes to these mountain dwellers and deals with these people to his own advantage, without having first discussed the matter with the eldest member of a family, then the eldest or the next in line goes immediately to this person and asks him: "In whose will are you doing this?" Should his reply be: "It is done in the will of the Great Spirit!" then he will not be interfered with in his undertakings.
8. But should the answer be: "In order for me to profit, this was absolutely necessary," then the eldest will give him the following lecture:

---

[34] People from the plains and valleys. – ED.

9. "Listen, my brother, in the Great Spirit! How is it possible that you have more than is necessary, which lies only in the will of the Great Spirit? Do you have an additional need which is apart or separate from the need in accordance with the will of the Great Spirit? Therefore, I advise you as a true brother in the Great Spirit to cease this work immediately so that you do not become hapless while in the midst of carrying out your intentions. Should you be destitute and homeless, look to our house, for there is enough room to accommodate hundreds like you. If you did this deed because of a hidden urge of selfishness, then fall instantly upon your countenance and implore the Great Spirit earnestly and remorsefully that He may spare you from a proper punishment. The Great Spirit is exceedingly benevolent to the good, but extremely strict and just towards those who are contrary to His Holy Will, which is above all!"

10. Following such an address, the foreigner who did not have the authority to commit these deeds will immediately abandon his intentions. Should he refuse, then the elder will reply: "Do as you please as far as I am concerned and may it be granted to you for all times of times that your sins do not become greater in the eyes of the Great Spirit. But be on guard that your punishment does not strike you in the open field."[35]

11. Upon which, the elder offers his hand and leaves and lets him continue with his work. When the elder arrives at his home, what does he do? You might be of the opinion that the elder will dispatch several members of his family, as is the custom on earth, with ropes and spears to take the offender into custody and take him home to

[35] When no one else is around. – ED.

dispense the proper punishment. That, of course, is not the case among the people of this planet, especially when it comes to the mountain dwellers. When such an occasion arises, the elder informs the members of his family as to the circumstances, and he requests that they all gather and beseech the Great Spirit with the supplication that He may be charitable and merciful towards this brother who has forsaken himself and who has acted contrary to the will of the Great Spirit, and that the Great Spirit may return the offender to the true cognizance, where a human being should have no other need than the sole will of the Great Spirit.

12. After the family members have prayed in earnest for a certain period of time, the eldest collects himself in his heart and calls for the Spirit of Light, who is always ready to dispense the proper advice. The elder asks the Spirit to proclaim to him the will of the Great Spirit for the permanent welfare of the particular lost brother. Under those circumstances the Spirit of Light never neglects to inform the elder as to what to do.

13. Should the offender be a hardened, willful and selfish person, then the elder is given the task of taking the foreigner into custody and leading him to the heights where the family dwelling is located. While there, the offender is given food and drink and is lectured in the cognizance of the only valid will of the Great Spirit, even if such a lecture lasts for seven days. Then the offender is led into the temple, and there he promises the Great Spirit the greatest obedience from the bottom of his heart and, consequently, never to take another step without the will of the Great Spirit.

14. If such an offender converts when this is carried out, then after a prayer of thanksgiving he is given a fair amount of different kinds of food supplies and is led by three people from the family to the place which he has indicated as his home. If it is a fact that they find

his dwelling where he said it would be, this ends the entire procedure, except that he is again reminded by the three in a brotherly but serious manner to abide by what he has promised; he is also admonished as to his gratitude towards the Great Spirit.

15. Should it so happen that this stranger's dwelling is too far away from the mountains or that he has no dwelling at all, which is mostly the case when it comes to the dwellers of the depths, in the first instance he is released at the foot of the mountain with a most urgent and extremely threatening admonition never to break his promise again. Then he is blessed and set free.

16. But if he is a vagabond and, even though when he was in the mountains he said he had a home, has in fact no place of his own, despite such a circumstance he is set free. A situation like this arises very rarely, and it is pointed out that he did not deceive the mountain dwellers, but only the One Whose will they fulfill at all times. They also make him aware that this is the greatest of all evils which a man can commit, and that he is now on his own and must manage to deal with the One Who knows all his thoughts before he thinks them.

17. They show the offender all the negative consequences which are based on experience and leave him without blessing him. Whosoever has committed an offense against them is blessed in order for that person to return to the Great Spirit; but nobody dares to bless a person who has committed a sacrilege against the Great Spirit unless it can clearly be seen that the offender is still in the grace of the Great Spirit. Should this be the case, he will be blessed.

18. If, as happens frequently, the offender is soon thereafter punished by the Great Spirit, the Saturnites ask the Great Spirit for forgiveness on an almost daily basis for the sacrilege he has committed against the Great Spirit; nobody dares to bless such a convict

unless they come into the knowledge spiritually or in a natural manner that the Great Spirit has begun to moderate the imposed punishment. That is the procedure with a hardened criminal.

19. If the person is not a hardened criminal, then the elder dispatches three messengers laden with fruit to the offender where he is still carrying out his work. As soon as they arrive, they instruct him to cease with his work immediately in the name of the Great Spirit, and they instruct him about the will of the Great Spirit, forgive him his deeds and take him into their midst and lead him down the mountain where he has stated that he lives.

20. There they make him a gift of fruit and say: "Brother, so that you do not commit any more sins against us in the future and, most importantly, against the most holy will of the Great Spirit, we let you go free, that you may visit us whenever you wish and nevermore go empty-handed to your domicile. We have learned from the will of the Great Spirit why this has to be done. But should you ever dare to sin again as you have done now, then you will suffer the punishment of the Great Spirit once you take the first wrong step."

21. Then they shake his hands, bless him, and admonish him to offer prayers of gratitude to the Great Spirit and return home.

22. These are the entire judicial procedures in such proceedings on the part of the people on Saturn. In the next chapter we shall pursue similar political constitutions and proceedings.

36

The Metal Industry And Other Crafts Or Trades True Socialism Of Altruism In A Trade And Commerce

1. Furthermore, the manufacture of the necessary metal tools that are required to hew and trim the trees, for the manufacture of

house utensils, to loosen the ground and for cutting their food and other things all belong to the lawful constitution of the people on Saturn.

2. Where are these tools manufactured? There are some factories on this planet, especially in the foothills of mountains, in which a useful metal similar to your iron on earth is turned into all kinds of different tools and utensils.

3. Who owns these factories? So that every neighboring family community has their rights to take these products in accordance to their needs from the factory, every neighboring family alternately sends workers to the factory to produce this metal under the direction of a factory elder. In the factory itself, the metal is not made into tools; instead it is only made malleable and hence prepared for further use as the bar-iron is produced and prepared for further use on earth.

4. Once a worker has worked in this factory for the required time of 100 days, and if, for instance, this factory has 100 workers, then the metal produced is divided into 100 parts. After the completion of the worker's term in the factory, every worker receives his rightful share, which is taken to his family dwelling as community property. The share a worker receives weighs anywhere from 1,000 to 1,500 tons in earthly terms.

5. What do the Saturnites do with this metal? If the family still has enough useable tools, then the metal is wrapped in foliage and the patriarch becomes the custodian of this metal. Should their tools require replacement because they are worn out and can no longer do a proper job, then they proceed under the direction of the patriarch to manufacture new tools.

6. How is that accomplished? Do you think that this is done with a regular fire, as on earth? The answer is no! It is accomplished in a

very peculiar manner and yet is a very simple procedure. The Saturnites make use of a pumpkin-like fruit which has a regular arch like a concave lens at the bottom surface, the diameter of which is anywhere from 120 to 180 feet. The outer bark of this pumpkin fruit, especially the bottom part, is arched-in, has a luster and is smooth like the finest polished steel; with this concave part of the fruit the Saturnites reflect the sunrays and direct the focal point onto the metal bars. It does not take more than a moment to make a large quantity of these metal bars white-hot.

7. Once that has been accomplished, whatever metal is required is cut off of these white-hot bars and forged with a metal hammer into some tool of which they are in need; this is done on a very smooth anvil which is as hard and as smooth as a diamond.

8. When it takes a blacksmith on earth half an hour to forge a sickle, in the same amount of time a Saturnite produces at least ten, even though a finished sickle weighs approximately 110 pounds in accordance with earthly measurements. If you take this into consideration, you can imagine how skillful a blacksmith on Saturn is in his art.

9. Here another question has to be answered: Who is skilled in this kind of work? This is not very difficult to answer, because among the people on Saturn it is part of their domestic constitution that every man be skilled in every required craft, so that no difference exists as far as their station in life is concerned and one craftsman cannot say to another: "I am of greater necessity than you are, and my products are more important than yours." Instead, everybody can do what his brother can do. Therefore one person can be of benefit to the other in everything that is required. And should one or the other be in succession to become the elder in the family, then he has the experience and the skills to lead and give the proper direction in all these matters.

10. When it comes to the necessities of life, since all of them are experienced as far as crafts are concerned, there is, so to speak, no horse-trading or profiteering, especially among the mountain dwellers. They also do not produce any surplus of products, so that they cannot sell or trade it to their neighbors; all products they produce are in accordance with their own domestic requirements.

11. However, regardless of that, should a neighbor be in need of something he does not have, because he does not have the same means as someone else, he is not asked: "What will you give me for what you require?" — the only question he is asked on an occasion like that is about the will of the Great Spirit. If he possesses this testimony, which is the only thing that is valid on Saturn and is the only valid currency, he will then be immediately given what he requires for his task and it will become his property, and nobody is allowed to think of any reimbursement.

12. This happens as a result of a political law among them, which is as follows: Who is more than the Great Spirit? What did we give the Great Spirit for the use of this great world which is abundantly furnished with countless goods? However, it is only proper that we thank the Great Spirit for any gift. If we would accept from our brother just one single thanks, how would we appear to the Great Spirit, if we demand from our brothers only what the Great Spirit is entitled to!? Therefore, woe to the one who would allow it to be thanked by his brother for a gift, where actually he should thank the Great Spirit for giving him the honor to serve his brother.

13. On account of this worthy reason, on Saturn a human being does not knowingly accept even the slightest thanks from his brother, much less would he accept any other reimbursement. Therefore all commerce between Saturnites is based on this premise.

14. On Saturn there are no exchange offices, no customs and excise offices, no taxes on goods, no inspector of goods; and profiteering is unknown on Saturn.

15. A tool which is in constant use on earth is a scale; this tool is completely foreign to the Saturnites. The only kind of scale they know is the sole will of the Great Spirit and the needs of their brothers.

16. A second tool, which on earth is called an ell,[36] is also foreign to the inhabitants of Saturn. Nothing is measured by the ell as it is on earth; the most infallible measure for the inhabitants of Saturn is the word of their brother in accordance with the will of the Great Spirit; with that they measure what a neighborly brother asks of them.

17. Commerce based on these principles would be a much better solution for all the inhabitants on earth than all the stock and money markets, banks, exchange offices, stores, bars and saloons. A somewhat better intellectual power would tell you: What did we give God for all the products on earth, and how much did we pay Him for the earth, since we rule on earth as if we were the immediate owners?

18. As aforementioned, if you had just a little more intelligence you would have to ask yourself the above questions out of necessity. All your actions will clearly tell you how unjust you are from My point of view on My property. I alone created and established equally for everybody to appropriate these products from My loving hands, and you jealously, greedily and violently process them for

[36] A measuring tool. – ED.

certain purposes and then sell them to your brother for an exorbitant price when he is in need or when he has a yearning for them.

19. But let us leave this crying shame on earth and turn again to Saturn, where human beings are still in possession of those treasures which are not attacked by rust and consumed by moths, and let us examine several more pages of their otherwise very small political law book, which is written in their hearts!

37
The Manufacture Of Woven Material
No Unnatural Utilization Of Natural Products
The Clothing Order

1. We have just perceived these human beings as blacksmiths, and learned how they "sell" their tools or products to a brother when the need arises; now we shall learn about the cloth makers.

2. When we were given the description of the plant and animal kingdoms, we learned that there are plants on Saturn that grow a kind of very long hair from the blossoms and also from the leaves; and that many animals are extraordinarily rich in wool and some of them have a remarkable abundance of long manes of hair; therefore it should make sense to you that the Saturnites make use of all these things.

3. How are these materials used? There is not much difference between you, the inhabitants of earth, and the inhabitants of Saturn. These materials are spun into threads, which are of course somewhat stronger than some of the ropes on earth. Nevertheless, they are fine enough in proportion for wearable fabrics to be woven from the yarn for these huge human beings.

4. Who spins and weaves this yarn? This work is done only by

the women, though not in the same manner as you weave your fabrics on weaving looms, but similar to your women when they knit stockings with knitting needles. That is how complete articles of clothing or garments are knitted on Saturn, namely with the help of two long sticks, always made from wood. The Saturnian women have such great dexterity that in one day a woman can produce a fabric of more than 250 feet long and 12-1/2 to 15 feet wide in accordance with your earth measurements.

5. Are these fabrics or materials dyed? On Saturn nobody dyes these materials. Here another domestic law applies, which because of so much vanity in the depths is as follows:

6. A human being is an offender if he wants to make something better, more beautiful and more perfect than the Great Spirit created it! Woe to those who want to make something red which the Great Spirit has given us in white! Woe to those who want to straighten out something which the Great Spirit has given us curved! Woe to those who want to make food more tasteful than the Great Spirit prepared it!

7. Whosoever acts contrary to the will of the Great Spirit in this matter will be looked upon by Him with resentment, and He will send one evil after another onto the physical body of that person in the same manner that He does it in the depths, where the human beings pay no attention as to how the Great Spirit has everything extremely well and wisely arranged. Therefore it is not necessary for a person to make changes. Instead he should gracefully accept how the mild hand of the Great Spirit gives it to him. Our only reason for being here is to use what the Great Spirit gives us, not to improve or beautify His gift so that we may use it.

8. There is one exception to this rule, and that is the metal which the Great Spirit placed raw or crude into the ground; that is why we

must bake it first before we can use it for our benefit. The reason we are allowed to do so is because the Great Spirit Himself taught us how to do that. In accordance with His will we can also soften some fruits on the fire for easier digestion, and we can hew tree branches for storage buildings, because all these things were taught to us by the Great Spirit.

9. But that we should give something another color or another polish, that He never taught us. It is therefore a great offense for the one who changes white into red, green into black and blue into yellow, or who dyes or colors something in the reverse order.

10. Amongst each other, we are one kind of brother and sister in the Great Spirit. In this respect there exists no difference — all of us were created by Him. Why then should we differ from each other in the color of our garments?

11. Why should the loincloth, with the belt that we wear which reaches down to the knees, be any other color but blue, because the wool that is used is blue by nature. The garment we wear on our upper body is red, like the mane of the animal from which it is made; our hats have the color of straw from which they are made, and our legs are covered with a garment that is green like the wool of the trees and plants from which it is made.

12. The women should also remain unchanged as far as their flowing blue shirts are concerned, and should always use as their outer garments the beautiful leaves of the dwelling tree (*gliuba*); and they can use many other things to adorn themselves with, such as what the Great Spirit grows for them on trees, shrubs and animals. The women in the mountains distance themselves from the excessive love of splendor and luxury more than the women who live at the banks of the great rivers and shores of the lakes, who find great pleasure in adorning their effeminate and tender bodies with all kinds of finery.

13. It is our duty on the holy mountains to be steadfast in everything and to be faithful to the will of the Great Spirit.

14. The manufacture of fabrics and the clothes themselves is one of the longest and most extensive house rules, as is the manner in which these clothes are to be worn.

15. The mountain dwellers are very generous with these fabrics, as they are with everything else. Should an almost naked person come into the vicinity of these mountain dwellers, his nakedness alone serves as indisputable evidence of the Great Spirit that anyone who has available clothing must clothe that naked person. Whoever refuses to do so is committing an offense like no other, and the punishment is banishment for up to three years, so that the person will learn during this isolation how painful it can be to wander about naked.

16. You might think: How can a person tear his clothes in one, two or three years to point of nakedness? Remember that one year on Saturn is almost thirty years on earth. If you consider this fact together with the idea of banishment for up to three years, it should become clear to you that in that length of time a garment will no longer look good or be in one piece when it is worn day and night.

38
The Stature Of A Man And A Woman
Procreation, Pregnancy And Birth
The Constitution Of The Children

1. We shall discuss in the chapters that follow other very noteworthy house rules. However, for now we shall familiarize ourselves with the stature of the Saturnites, the male as well as the female. What does a woman on Saturn look like?

2. By now you know that all things on Saturn are more exalted, more magnificent and more beautiful than on earth, and that applies very much to the human beings who inhabit Saturn.

3. The women on Saturn have perfect figures and are beautiful far beyond your earthly concepts. Even though she would be considered a giant on earth as far as size is concerned, her beauty would compel you to admire her. Her body has the most wonderful curves throughout. Not one part of it is disproportionate. Her whole being is softness; nowhere will you find any hardness.

4. In the sun her skin is as white as snow. Only in those areas where the skin is most delicate does it appear as a pale red. Despite the size of the Saturnian women, their skin is lighter, much finer and more delicate than the skin of the most delicate ladies on earth.

5. Their fingernails and toenails are different from those of earthly women. The color of the nails is as if someone were to gently cover polished gold with a crimson color. However, where the nail grows past the fingers and toes, the nails assume the colors of the rainbow. Therefore the women on Saturn are adorned by nature more beautifully than the women on earth would be even while wearing the most beautiful gold rings.

6. The bosom of a woman on Saturn is elastic and developed to perfection, and the color of the skin has a reddish tinge; at the same time it is so truly ethereally delicate that even the bosom of the most delicate ladies on earth would be like a hard pebble by comparison.

7. The neck is neither too long nor too short, but rises above the shoulders in the most beautiful and harmonious proportion to the body. The neck is adorned with a truly heavenly, beautiful head; any exception to this is a rarity.

8. The forehead is of medium height and dazzling white. The

nose is straight, sublimely soft and gentle in all its parts. The eyes are mostly exceptionally large, the pupil is pitch black, the iris completely azure blue, and the remainder of the eyeball extremely white. The eyebrows of every woman are thick and of a dark gold-blond color. The hair of the head is extremely soft and in most instances reaches the knees and is of the same color as the eyebrows. The mouth is proportionately small; the lips are crimson red. Behind these beautiful lips the upper and lower jaws are adorned with the most beautiful pearly teeth. The chin protrudes a little and is slightly reddened, just like the cheeks. The ears are small rather than large in relationship to the head and are also slightly reddish in color.

9. This completes the description of the stature of a woman on Saturn. Do not assume now that in a physiognomical respect one woman resembles the other. On Saturn the same rules applies as on earth — you will not, among 10,000 faces, find two that look alike.

10. Now if you can imagine the figure described wearing the clothes described in the previous chapter, you should have an idea what such a woman looks like. This completes the description of the Saturnian woman; we shall now turn to the Saturnian man.

11. What does a Saturnian man look like? As you already know, a man on Saturn is considerably taller than a woman. That is something that does not occur too often on earth. On the average a man on Saturn is usually 15 to 20 feet taller than a woman. Initially you will probably find such a ratio a little peculiar, when you take into account that on Saturn a woman, when compared to a man, is proportionately the same as a ten-to-twelve-year-old girl compared to a fully grown man.

12. But when you find out that the act of procreation between two Saturnites is quite different from procreation on earth, then you will

agree that the variance in proportions is quite proper. Since we have touched on procreation, we will say a few words about it before we continue with the shape of the Saturnian man.

13. How do Saturnites procreate? Through the sole love and the firm will, exactly as it could have occurred on earth if only man had not fallen away from Me before I could bless him.

14. When a man on Saturn wishes to procreate, he presents himself with his wedded wife to the eldest (a man on Saturn never has more than one wife). The eldest blesses him in the name of the Great Spirit, upon which the man and his wife fall to the ground and fervently ask the Great Spirit for the awakening of a living fruit.

15. When they have completed their supplication, the man takes his wife into his arms, presses her against his heart and gives her one kiss on the forehead, one kiss on the mouth, and one kiss on the bosom, whereupon he places his right hand on her body and fixates her with his will. This is the complete procedure of procreation, during which the man and woman both perceive a truly heavenly, pure love which inspires them and makes them overly cheerful for the longest time.

16. Once the procreation has occurred, both spouses fall to the ground again and thank the Great Spirit for it, and at the same time they ask Him to bless the fruit so that it may prosper. Upon which they present themselves to the eldest again, who will bless the wife; and after that her husband blesses her.

17. On Saturn a pregnancy has a duration of only three months, and the woman has no external signs other than a more lively coloration of her bosom.

18. The birth is without pain. At birth the children are very small, about the size of a five-year-old child on earth. However, they grow very quickly and in three earth years are anywhere from 72 to 90 feet tall.

19. In the first years the children are so light that they float in the air like a feather fluff; they become specifically heavier when they are weaned from the mother's breast and begin to eat stronger and firmer food; that is also the reason why on Saturn you never hear that a child has sustained any injuries by falling from some heights.

20. By taking everything which has been said into consideration, it is easy to understand why a woman is smaller and weaker than a man by almost one-third.

21. A man on Saturn is a complete image of Myself. His height gives him the certificate that he is the lord of nature on this planet. His stature signifies that the proper form of a man is one which resembles the hardness of a rock, without any of the softness of a woman's bosom. He should be in every aspect the image of the One Who created him, containing within himself all the perfection of energy, might, strength, firmness, will and the magnificence and beauty of all forms.

22. Therefore, if you would like to imagine the form of a man on Saturn, then you have to imagine a perfect youth on earth, one upon whom the muscles still show some kind of continuity and not much separation. Transfer this perfect male human form to a Saturnian man, of course on a much larger scale, and you have his stature clearly before you. However, a man on Saturn has considerably finer skin than a man on earth.

23. The chin of a Saturnian man is covered by a proportionately large beard and curly, mostly light blonde hair on the head, flowing down to the waist. It should be mentioned that the beard and especially the eyebrows are always a little darker than the hair of the head.

24. All parts of his body are in a most beautiful proportion; some parts, such as the cheeks, chest and arms are a little more reddish than those of the woman.

25. Secretly you might wonder: Do the Saturnites have no genitals similar to the ones people have on earth? Yes they do, but in proportion they are not as large and peculiar, because on Saturn they only serve one purpose.
26. Imagine such a Saturnian man in all his magnitude and true human nobility; you must grant him the credence that his form represents all the nobility, all the dignity and all the magnificence which is the primary source of all things.
27. Even though the woman is also very lovely on account of her curves and softness, only the man is truly beautiful and remains eternally beautiful and completely perfect in everything.
28. All the inhabitants on Saturn are fully aware of this, and that is why they thank the Great Spirit at all times for the noble form which resembles the perfect form of the Great Spirit, which the people on earth would never think to do!

39

The Cognizance Of God By The Saturnites
More Respect Than Love
Knowledge Of God's Incarnation On Earth

1. How do the Saturnites know that their form is the image of the Great Spirit? They know because the Great Spirit Himself has revealed Himself to them many times. And there is rarely a family elder who has not seen the Great Spirit at least once.
2. Therefore they know God as a perfect human being. Therefore, the highest of all principles of the teachings of the Great Spirit is as follows:
3. "God, Who is the Great Spirit, is the most perfect human being of all. He has hands and feet like we have, He has a body like

we have and His head resembles ours. But He does not labor with His hands and He does not walk with His feet; instead all His infinite might lies in His will. And with the incomprehensible energy of His wisdom He creates and guides all things."

4. And since the Saturnites have the proper concept of Me, that is why they recognize themselves much more easily and much sooner, and they are fully aware in their spirit that they are not fleeting images, but eternally remaining images in the spirit of the One after Whom they were formed and created.

5. Do the inhabitants of Saturn love the Great Spirit? Yes, they do. But their love consists mostly of immense respect or awe rather than an endeavor to get closer and closer to the Great Spirit so as eventually to become completely one with Him.

6. They also know very well, and are informed about it by the spirits, that the Great Spirit, on a small celestial body close to the sun, became a human being with flesh and blood; and that the human beings on that celestial body failed to realize Who He was and that He was physically killed by them — all of this they know very well.

7. However, they have great difficulties in understanding how it was possible that these human beings failed to recognize Him. And they question the spirits that come to them as to what the people on this celestial body are doing now and if they still have not recognized the Great Spirit.

8. When they receive mostly a "no" to these questions, they become quite sad and pray often and ardently in their temples that the people of such a highly blessed celestial body may recognize, just for once, the One Who bestowed upon them such a blessing; whereas when the Saturnites just think in the slightest about His greatness they begin to tremble throughout. After holding their

breath for a considerable length of time, they say with very strong words:

9. "Oh, if we would have been honored with this blessing, if the Great Spirit would have clothed Himself on our planet with our flesh and blood, truly we would shine more than a thousand stars on top of each other!"

10. That is how the Saturnites express themselves when they hear something about the earth. They have a great yearning to see the earth. This cannot be achieved in physical form, but there is hardly a spirit of the Saturnites who does not visit the earth as soon as he has discarded his physical shell.

11. But since such a spirit cannot see the material, he views the spiritual earth and, through the correspondences, he can see the material world from the spiritual world. Once he begins to recognize the human beings on earth, he becomes very sad and soon leaves this celestial body.

12. In due course, when we discuss the religion of the Saturnites, we shall learn more. But for now, we shall return to some of the political constitutions.

40

Law Against Arrogance
The Law Of Cleanliness
Aversion To Everything Dead
Funeral Services
Marriage

1. What does this political constitution contain? Its content is as follows: Nobody is allowed to talk to any great extent about his physical beauty and size or about any other aspect of himself.

2. So that this important commandment is obeyed at all times, they already impress upon their children that they are very small and all worldly greatness appears to the Great Spirit as a mere nothing. That is why there is not one patriarch or elder and much less a family member who thinks anything great about himself.

3. They make the following comment regarding the beauty of the body: "All of us are equally beautiful as the image of the Great Spirit. Anyone who would say and believe that he is beautiful by himself and therefore have an advantage over all the others would no longer resemble the eternal primary image, and that would make him uglier than the ugliest animal on the celestial body."

4. The spirits of the departed assist in maintaining this law. Should someone be beset by vanity, he will soon be confronted by some horrible, distorted, smirking face. Whoever has been punished once in this manner will soon discard all his vanity; the Saturnites know that the spirits do not jest, because it does not matter in what manner they express themselves, they are always serious. This is a political law which is observed by the important and the less important, by old and young.

5. As far as this greatness is concerned, it remains for always in the spirit, namely that you consider yourself as small or as the least important person. That is why the spirits on Saturn do not have a good relationship with the spirits on earth, because nothing is more predominant than their putative greatness.

6. The next law is a domestic law which is based upon the recommendation and maintenance of cleanliness; that is why you will rarely find anyone on Saturn who is not clean, whether as regards body or clothing.

7. Regardless of the law mentioned above, their main teachings are not to attach too much value to the physical body, because it is

mortal; however, the opposite applies to the spirit, because he is immortal.

8. This is why the Saturnites have an aversion for everything dead and, as you know, they will not live in dead houses, but only in live ones. Much less will they tolerate anything dead in a temple which is blessed by God.

9. Everything must be kept clean, especially the body, since it is the dwelling place of the immortal spirit. And that is another house rule!

10. What happens on Saturn with the bodies of departed human beings? They are not buried as on earth, nor are they cremated as is the custom in some earthly countries. The bodies are taken to a place where there is a pyramid tree forest. While there, the dead are placed face down on the ground and covered with the branches of the pyramid tree. The bodies of dead females are hung from a branch close to the trunk of the tree by tying their feet together with the head almost touching the ground.

11. You probably might think that a bad odor would pervade the air for quite a distance when such large bodies begin to putrefy, but on this planet this is not at all the case; instead, it is the opposite. Since the bodies of the Saturnites are more ethereal and lighter than your physical, coarse "serpent bodies," they evaporate in a very short time after the person has departed, and this evaporation produces a very pleasant fragrance in that region.

12. When this fragrance enters the nostrils of a Saturnite, he falls to the ground full of gratitude towards the Great Spirit and implores Him to permit the spirit of the one whose fragrance has touched his nostrils to come to him so that together they can give praise to the Great Spirit for the deliverance from the prison of the flesh. This happens all the time, especially when the petitioner is serious.

13. Do the Saturnites mourn when someone experiences physical death? The answer is no. For example when an elder has died, the next oldest in the order becomes his successor, and asks all family members to fall to the ground and thank the Great Spirit for bestowing upon the patriarch such great blessings that He has called him into the eternal life.

14. After that they have to beseech the Great Spirit that He may graciously permit the spirit of the departed to appear soon to the new elder to lead him into the inner sanctum of the temple to bless him there for the sublime office of the Great Spirit.

15. This always happens visibly for the entire family to see. The spirit of the departed elder appears in his glory and asks the new elder in perceptible words to follow him into the inner sanctum of the temple and for the family to follow him into the general area of the temple.

16. While there, the spirit places the new elder onto the pulpit in the presence of the family, blesses him, and points out that the Great Spirit is pleased that the new elder has taken over the holy office and therefore the family should obey him in all matters and observe his words, whatever they might be.

17. He then asks the men to remove his dead body in the usual manner; he blesses the entire family and promises that the newly installed patriarch will be the head of the family as long as it pleases the Great Spirit, and he will remain the teacher and leader of the entire family in accordance with the will of the Great Spirit.

18. Following that, the spirit vanishes. The new elder and the entire family fall to the ground and thank the Great Spirit for everything that has occurred. Once the prayer of thanksgiving is complete, they all arise, immediately remove the body of the departed elder, and take him to a previously designated place and then go home in silence.

19. When a woman dies, they also ask for the appearance of her spirit. After her appearance, they pray a thanksgiving prayer at home. Then the body of the departed is taken to a previously designated place. The body of a woman evaporates considerably faster than that of a man; under favorable circumstances it evaporates so quickly that on the tenth day nothing remains, not even the bones.

20. The quick decomposition is accelerated naturally by the great needle tree (*uhurba*) which robs the corpse located beneath it of all its electricity with its millions of tips. As soon as the electricity has completely escaped from any natural body, the body vanishes as if consumed by a fire.

21. The observance of these rules is another one of those "important house rules" which are to be followed exactly and strictly at all times.

22. Is there another house law which is important and noteworthy? Yes! It is the law of the conjugal union of man and wife.

23. Through this law every man is strictly bound when he has reached the proper age for marriage, anywhere from thirty to forty, to take a wife of his choosing and satisfaction.

24. However, he himself cannot make it known to the one he has chosen for his wife. This can only be done through the elder. The elder calls upon the parents of the intended bride and lets them know the wish and the will of the suitor. Such a declaration is accepted with the greatest of gratitude and is considered to be the will of the Great Spirit. And such a suitor is never turned down, as is often the case on earth.

25. Then the elder takes the bridegroom to the bride, takes her right hand and his right hand, and joins them. They have to keep their hands joined and follow the elder into the temple, but only up

to the inner sanctum. There they lean with their faces on the cone-shaped altar, while the elder in the inner sanctum calls in prayer for the Spirit of Light.

26. On this occasion, as soon as the Spirit of Light with his veiled countenance appears, the elder asks the bridegroom and bride to rise. The elder then informs them of their conjugal duties, which consist of all the house rules which we have already learned, as well as some others which we have yet to learn about.

27. Once he has done that, the elder comes down from the pulpit and makes a gesture, as if he wants to separate the hands of the bride and bridegroom. But this is an old well-known rule, which is only symbolic and means that while they are in this world they should never separate, no matter what happens.

28. After this ceremony the elder steps aside, the Spirit of Light unveils his countenance, blesses the couple, approaches them and separates their hands. The significance of this is that the only valid reason for their separation is death or the separation of the spirit from the body.

29. Whereupon the Spirit vanishes and the couple are man and wife.

30. Now they offer a prayer of thanks to the Great Spirit. They beseech Him to bless the married couple with descendants that are pleasing to Him and to guide them in accordance with His most holy will. Once they complete this prayer the elder and the married couple rise and walk home with the greatest veneration, where usually a meal for the glory of the Great Spirit with the entire family present awaits them.

31. The next day the newly married couple is given a choice to either remain with their present family community, or to move elsewhere and build their dwelling. Should they agree to remain, a

branch is prepared as their dwelling and also a new house, a new kitchen and a storage building is built for them. Should they decide, however, to leave their family because of limited space, they are provided with all kinds of supplies and they can take their parents and other close relatives with them.

32. It has already been explained what happens when they find an unoccupied dwelling.

33. That is also a family house rule! We will, however, discuss a few more house rules and then we shall discuss their spiritual religious constitution.

41

The Inhabitants Of The Lowlands Visit The Mountains Their Reception By The Mountain Dwellers Half-Heathens And Their Conversion Courting In The Lowlands

1. It happens occasionally that families from the valleys and plains, because of corporeal health, travel to the mountains. If this is the case, the mountain dwellers have a rule whereby they receive these "seekers of health" warmly, and provide or obtain for them whatever they deem necessary for their health.

2. Should these seekers of health decide to dwell in the mountains permanently, then the elder of the mountain family appoints a leader from that family to help the people of the lowlands who wish to settle. Should the people from the lowlands want to retain this person as their leader in the future, he is obligated to honor their wish.

3. Should this leader be married, he leaves the new settlers only for as long as he needs to get his wife and children from his home,

and to receive on this occasion from his elder the blessing for his new office. Once he has put everything in order, he sets out with his wife and child and with whatever other supplies he requires to the newly settled family.

4. Once there, he begins to instruct this family in all laws and customs of the mountain dwellers and encourages them to build a temple and also in accordance with numbers and needs to build a storage building and the living residential homes, because these are also laws which the mountain dwellers have to obey!

5. It also happens that these valley and plain dwellers travel to the mountains strictly for the beautiful view. When the travelers meet a mountain family, they are stopped and questioned in a very polite manner regarding their intentions. Usually they answer these questions in a very polite manner; since, as a rule, their intention is to climb the summit of a high mountain because of the spectacular view, the elder of the mountain dwellers will inform them which of these summits do not present any danger to the climbers. Should any of the mountain peaks pose any danger, then these holiday makers are politely prevented from embarking on this endeavor, and informed in detail as to the dangers they may face if they do not change their plans. Usually the advice is taken and a change of plans is made, upon which the travelers return to their home.

6. If, however, these mountain peaks are accessible without posing any danger to the climbers, then the mountain dwellers will provide a guide for these travelers. This guide is responsible for three things: first, he must lead this group to the summit on the best and safest trails; second, he must supply them with food and drink, which is usually carried by the house servant *fur*; and thirdly, he must explain everything to them, and at the same draw their attention towards the Great Spirit in everything he does.

7. The guide cannot ask anything in return for his efforts, except that, after they invigorate themselves on the grandeur of the Great Spirit, he can ask the group for their assurance to follow the Great Spirit at all times and never to undertake anything without His expressive will.

8. Once he has obtained this assurance, the guide reminds the group that they have to thank the Great Spirit for everything they have learned and experienced, and, at the same time, they should ask Him to let them reach their homeland safely. Once that is done, they set out for home.

9. The group is taken to the guide's home for food and drink. Once they have properly refreshed themselves, they are politely reminded to give thanks to the great and holy donor; they are also greeted and blessed by the elder and set out to return home to the lowlands.

10. In some of the mountain crannies live the "pointed-foot flesh doctors"; the plains people are forewarned about these charlatans by the mountain dwellers. It is pointed out that these people are trespassing fugitives from the lowlands who have settled in the mountain crannies for sheer greed, by cheating inhabitants of the lowlands with all kinds of worthless and ineffective things. The mountain dwellers warn these gullible people from the lowlands in the following manner:

11. "The Great Spirit has set thousands of wholesome and healing herbs and fruits into the ground for the maintenance of our bodily health, so that they grow to strengthen us. He created clean water throughout the entire world and planted giant trees in the soil to attract all the toxic vapors and change them into a fragrant, pure air. And the Lord Who is exceedingly benevolent in all His infinite might placed a magnificent sun in the firmament, whose rays

awaken the most wholesome energy in the star lilies which are so wholesome that their power reaches as far as the rays of the sun. And that is how the Great Master has arranged the firmament with the light white ribbon[37] which illuminates the night for us so delightfully after the sun has set that it is almost as pleasant as the day, and even during the night we are invigorated by the strong rays of this ribbon. Besides this ribbon, the great Workmaster has placed seven great lights,[38] of which several alternately delight us at night, even at the time of the full shadow.

12. "The Great Spirit has been exceedingly benevolent and has graciously cared for us all. And we, the inhabitants of the mountains, have experienced this at all times. That this is so is because we have never been afflicted by an ailment. When someone dies, it is never of an ailment; they only die as a result of their completely matured spirit, which for eternity will never die again, but live forever. We are witnesses of this existence after death. We can prove that this is so at any time right before your eyes.

13. "It is therefore great folly to attempt to obtain a remedy to provide you with eternal life from a human being whose eyes are full of deceit, since he himself is far removed from the life of the Eternal Spirit. However, the remedies with which the Great Spirit has blessed us so abundantly are considered to be inadequate according to these pointed-foot doctors and not fit for use.

14. "Dear brothers from the lowlands, I will tell you that in the

---

[37] Another name for Saturn's rings. – ED.

[38] Saturn's moons. – ED.

future you should no longer seek your well-being from these quacks; instead, seek your well-being seriously in the will of the Great Spirit, for then you will remain healthy until your spirit has fully matured.

15. "When your spirit has fully matured, which means when it has completely become a lord of life out of the will of the Great Spirit, you will nevermore taste death; you will be able to step out of your flesh and blood without pain and without restraint with the clearest and fullest consciousness, and you will discard this heavy garment of flesh and bones with the greatest of gratitude towards the Great Spirit."

16. After the lecture, these so-called long-life remedy seekers are released with blessings to return to their homeland with a much better life remedy.

17. In some places on the plains and in the valleys, especially along the lakes and rivers, there are certain heathens who consider the white ring in the firmament to be a divinity; others consider the ring to be the path upon which the Great Spirit walks to and fro and from whence He looks down upon the planet to see what the people are doing. These half-heathens often climb the mountains in the belief that they are closer to the white ring and perhaps might even reach the ring itself.

18. The mountain dwellers feel a kind of duty of love to lead these travelers onto the right path and to show them what the ring and the moons really are and what their purpose is. They accomplish this feat through their firm will by placing these lost human beings into a state of clairvoyance in which these travelers are able to examine the ring and the moons from beginning to end.

19. Once these heathens have experienced this, they realize how wrong they were, whereupon they are instructed in an affectionate but wise and serious manner in the true cognizance of the Great

Spirit and His will. It is left to their discretion whether they wish to spend the rest of their lives in the mountains or whether to return to the valleys and plains where they lived before.

20. Should they decide to stay in the mountains, then they are provided with living quarters. However, should they prefer to live in the lowlands, they are blessed and provided with food and drink and sent away in the name of the Great Spirit.

21. But as long as these heathens from the lowlands do not give up their erring ways, there is no hope that the strong mountain dwellers will let them go. Should any of these heathens be of an obstinate or rebellious mind-set and take the initiative to escape or attempt to harm someone, their actions will be to no avail because at first there will be threats, and then a punitive angel-spirit will be present who, with the simplest of means, makes such an obstinate and rebellious person understand how little he will accomplish with his defiance. After such a lecture, the obstinate person usually mends his ways. Should he persist with his obstinacy, it happens that he is punished by the avenging angel, either with severe abdominal pains or, under extraordinary circumstances, even with corporeal annihilation. The effect of this punishment is that such a person will remain for quite a while a guard of the night and coldness.

22. One of the most important house rules for the mountain dwellers is the following: Should the majority in a family be male rather than female, then the male majority is at liberty to look for a wife among any other neighbors. Should he not find a wife there, he can go to the lowlands. Should he not be able to find a wife there, he can travel even farther. It does happen that at times a bridegroom travels to two, three, or even four continents in search of a wife.

23. Should he find a wife in these far away places, as a rule there

is no objection if there exists a female majority. The bridegroom is at liberty to remain there with his wife, and this is usually the case; or he can return with his wife to his homeland, but he is obligated to visit the house of her parents every three years as long as the parents are alive. Once they are no longer alive, then the obligation has been fulfilled.

24. Should the majority of the members of a family be female, then messengers are sent to inform the neighboring families; on this occasion one or several suitors may be considered.

25. The elder calls upon the Spirit of Light to indicate to him the will of the Great Spirit regarding which of the suitors is the most worthy among them; this takes place immediately. The one who has been chosen to be the bridegroom then takes his bride home after the wedding ceremony which was described in the previous chapter.

26. Should there be more than one eligible woman in one family and only one suitor, then the elder calls upon the Spirit of Light to indicate the one most worthy for that man. Whereupon those two are married in the ceremony described and the bridegroom takes his bride home.

27. Should there be the same amount of eligible women as there are suitors, then every suitor has the right to choose. Once they have chosen, the suitors have to inform the elder of their choices. The elder then confers with the Spirit of Light as to whether these choices are just, with the approval of the Great Spirit. If this is confirmed, the marriage ceremony is carried out without any further ado. Should the Spirit of Light not sanction the choices, then it is asked by the elder to indicate the proper choices in the name of the Great Spirit; this is done immediately, and the couples agree with great gratitude in their hearts. Upon which the couples are immediately married and the bridegroom can take his bride home. It is

obvious that the details of the ceremony which have been explained in the previous chapter, inclusive of all the customs, apply here as well.

28. Besides these house rules there are a few other unimportant household rules which will not be mentioned, because to know them would not serve any purpose. In the next chapter we shall discuss the religion of the Saturnites.

42

The Innermost Spiritual Religion Of The Saturnites
The Significance Of The Number Seven
Sunday Celebration
The Baptism Or Blessing Of A Newborn Child
The Temple Meal
The Sermon Of The Elder Supported By Spiritual Visualization
The Wisdom Of The Saturnites

1. The religion of the Saturnites has very little external ceremonial pomp and circumstance; instead, it is more internal and spiritual.

2. The ceremonial part, as you know, consists of a well-ordered and living temple, in which prayers of thanksgiving and supplication are offered to the Great Spirit regarding all important matters.

3. The holy number for the Saturnites is 7, and every number which is divisible by 7, such as 14, 21 and so on are all holy numbers. That is why a period of seven days, in other words the seventh day, is celebrated as a holiday.

4. The observance of this holiday forms the second ceremonial part; on this day all religious ceremony pertaining to the holiday takes place.

5. The first kind of ceremony is already known to you. The ceremony of the holiday consists therein that before sunrise all members of the family go to the temple; all the men walk in front while all the women follow behind. In the temple, the men stand on the right side while the women stand on the left. When they have all gathered in the temple, the elder leads the congregation in prayer until the sun rises, offering praise to the Great Spirit and thanking Him for all the blessings they have received. These prayers are always prayed with the deepest emotions in their hearts.

6. When the sun rises, everybody leaves the temple and enjoys the view of the coming day and the extremely beautiful panorama of this celestial body. When the sun is very high on the horizon, they return to the temple and thank the Great Spirit for the return of the day.

7. And when a woman has given birth, the newborn child is brought to the edge of the inner sanctum. There the elder places his hands upon the child and speaks the following words over the newborn baby:

8. "You came into this world as a weak guest, bound in all your powers in accordance with the will of the Great Spirit, Who is holy beyond all holiness and has power over all might, Who has control over all power, and Who is exceedingly faithful and steadfast in each of His words and in all of His promises; only He is the perfect and most highest Lord over all things which fill this earth and the entire infinite firmament. Therefore it is His will, as He is holy and holy beyond all holiness. That is how you should live on this earth until the end, completely in accordance with the will of the One through Whom you entered into this world, and then as a man (or in the case of a girl as a faithful wife) leave this world in all the truthful dignity and magnificence of the most consummate virtue!

9. "That is why I bless you here in the inner sanctum in the name of the Great Spirit, Who created and blessed you, your parents and me. Grow up in this blessing and multiply these blessings within you by strictly following the most holy and highest of wills! This shall occur at all times, now and eternally!

10. "As small as you are now, remain always that small before the Great Spirit, before us, your fathers and brothers and before your own self! And this has to be at all times in this life and the life beyond! Amen!"

11. After these words have been spoken by the elder, he insufflates the child, then the elder lets the parents bless their child and take it home. On such a holiday the parents are not obligated to return to the temple; instead they can remain at home and care for their child that has now been blessed. But if they prefer, they can remain in the temple.

12. Should there be no newly born child, then instead of the child's blessing the people will immediately begin with their morning meal in the temple. The Saturnites take their morning food, as well as the midday and evening food, into the temple right away early in the morning. It does not require any further explanation that they give a prayer of thanks to the Great Spirit before and after their meals.

13. After the morning meal, the elder ascends to the pulpit and addresses the moderately large family community, which in the mountains does not often exceed one hundred people. It is different in the lowlands, where such a family often counts several thousands.

14. What does the elder talk about in the temple? He is never short of subject matter. In this instance, as on all other occasions, a spirit stands by his side at all times and conveys to him what is to be said.

15. Usually these kinds of lectures concern the wonderful guidance of the Great Spirit, how He placed the human race from the dawn of history on this celestial body, and how the Great Spirit has led them to the present times in accordance with His wisest and holiest of all wills. On this occasion the elder often tells a story of times of yore. At times the elder explains the structure of their world, and sometimes the structure of the ring or the moons. At other times he might tell them about one of the different constellations, and he shows his listeners the guidance of the Great Spirit on those celestial bodies, and on these occasions he might occasionally mention earth.

16. But as soon as the elder mentions the earth, all listeners fall immediately on their countenances — not because they are in awe of this planet, but because they hear about the infinite love of the Great Spirit, and how the inhabitants of this earth address and call Him Father, which, for the Saturnites, is something indescribably holy, so that they always fall into a shudder of humility, especially when the elder reminds them of the ingratitude of the inhabitants of this planet.

17. On other occasions he gives them information regarding the spiritual world and life in the heavens.

18. After each of these lectures, especially when the subject matter has been the structure of their world, the ring, the moons and other planetary constellations, the elder transposes his listeners, sometimes many and sometimes just a few, into an internal visualization wherein they can view all of this as if they were physically present.

19. That is why the Saturnites, especially those who live in the mountains, are exceedingly wise and are enriched with considerable knowledge. The greatest scholars on earth would soon realize how insignificant their knowledge is if they were to enter into a scientific discussion with the least knowledgeable human on Saturn.

20. They not only know their own celestial body in the minutest detail, in as far as it is necessary and beneficial, but they also know other celestial bodies better than you on earth know the islands in your oceans. The Saturnites not only know the history of their own planet well, but they know the history of several other worlds.
21. No language is foreign to them. That is why they understand the spirits, no matter from which celestial body they come (since every spirit takes into the beyond, to a greater or lesser degree, the characteristics of the language of the world where he walked in the body). It is this understanding of languages that the spirits of your earth do not possess until they are completely born again in the spirit and are made suitable for heaven.
22. It happens often that the spirits of earth meet with the spirits of Saturn after their physical death, especially when they have this longing. The spirits from Saturn understand the spirits from earth immediately. However, that is not the case the other way around, especially when they are immature spirits from earth. Also, the spirits from earth do not see the spirits from Saturn unless they want to be seen by them. The reason for this superiority lies within the great and true inner wisdom of the Saturnian spirits.
23. Those are the fruits of the instructions and teachings of the preacher in the temple after the morning meal.

43

Profound Spiritual Observations Of Nature Communication With Angels And The Lord Himself

1. What happens after such a sermon?
2. The people thank the Great Spirit for such enlightenment

from their elder. The elder also thanks the Great Spirit and blesses all the listeners after they have offered the prayer of thanksgiving. Then they all leave the temple and go for hikes or walks together in the beautiful hills. They have conversations about the sermon they just heard, but they also discuss observations which they make about some object in nature that comes to their attention during these outings. All of them are inspired by a great friendliness and a mutual participation in everything that one or another may find and the comments one or the other may have regarding the admirable praise for the Great Spirit.

3. For example, one person will draw the attention of another to the structure of a flower, another to the movement of a cloud, a third to an animal or the flight of a bird, while yet another draws the attention of his neighbor to a bird that is singing and another discovers the glimmer of a distant lake or river somewhere. There are countless objects to which the Saturnites pay attention on such occasions, and they look upon these objects as at one time the man after My heart did when he observed My works.

4. At this point, I am telling you too: Whosoever observes My works with humble eyes and respects them will always find great enjoyment in them. But whoever examines My works with critical and erudite eyes is better off to remain in his idleness rather than to examine My works with his unconsecrated, profane eyes, like a gall fly which pierces into an acorn to place its pernicious brood into it. This fruit, when it comes to its bad maturity, cannot be used for anything other than the preparation of a black juice; this juice can be used for the blackening of a white or light surface, but it can no longer be used for cleansing what has become dirty or black.

5. But let us return now to our promenading people. How long does such a stroll take? Until midday. They all gather again in the

outer courts of the temple. There they offer thanks to the Great Spirit and have their midday meal in the outer courts.

6. Once the meal is over, they again offer thanks to the Great Spirit and some remain there while others stroll on the beautiful walkways around the temple and enjoy the multifarious splendor of the magnificent flowers which are planted in the beautiful flower beds in the outer courtyards and along the walkways of the temple. The women caress their husbands and children and tell them with their lovely voices what they have perceived and gained from the sermon of the elder as well as what they have learned during their stroll of what is good and true, beautiful and noble.

7. On such occasions, spirits and angels sometimes join them and discuss any matter concerning the Lord with anyone.

8. Sometimes the Lord Himself appears to them on such occasions in the form of an angel. As long as the Lord is among them and converses with one or another person, not even the elder is aware that it is the Lord Himself. But when He is about to leave such a gathering, He discloses His identity just before He escapes their visibility. But then He leaves immediately, because the Saturnites could not bear it if the Lord were to remain any longer with the knowledge that He is the Lord, since their respect for Him is so tremendous that not one of them would dare to speak His name while standing upright. Through which the commandment (which is not given to the Saturnites literally): "Thou shalt not take the name of the Lord thy God in vain" is obeyed in the most accurate, exact and most holy manner.

9. How long do these discussions in the outer court last? Until sunset. After that they all go again into the temple and glorify the Great Spirit and also sing praises to Him.

10. The elder ascends to the pulpit again and points out the great

blessings of the day, blesses them and concludes the ceremony of the holiday or feast day. After that they all go home in a thankful and merry mood.

11. Should something be left over from the midday meal, they distribute it among themselves and eat it thankfully. If nothing is left over, then they have their evening meal at home. They offer thanks to the Lord in a general way and the holiday is over and with it all ceremony, which in actuality consists of nothing but what has just been described.

12. This also concludes our day. The next chapter deals with the spiritual aspect of the religion.

44
The Spiritual Aspect Of The Saturnian Religion
Spiritual Rebirth Through Teachings And Practice

1. Since we have learned about the ceremonial part of the religion of the Saturnites, we shall now turn, as aforementioned, to its spiritual part.

2. Since we have sufficiently observed the ceremonial part, by now it must have crossed your mind that the ceremony where the angel-spirits of the heavens and quite often God Himself visibly appear and have discussions with the Saturnites is anyhow as spiritual as it can possibly be. Where could there possibly be something more spiritual?

3. I say: Let it be! The following will teach you how in something spiritual there is even more spiritually contained.

4. So that you understand this to some degree I will show you an example of how this is possible in accordance with nature.

5. For example: Take a pitcher of very good wine! Who of you will not understand soon and also feel from the effects that this wine contains something which is even more of a primary spiritual substance than the wine itself? However, just ask the next best pharmacist and he will tell you that from this wine you can distill the finest spirit (ether) of wine, and this spirit can be rectified several times so that the spirit becomes so volatile that, if one drop is poured from its receptacle, it will completely evaporate after it has fallen no more than 3 feet before it reaches the ground. Do you now realize how much more spirituality is contained in the wine, which is already exceedingly spiritual?

6. If this makes itself known in visible nature, by how much more will it reveal itself in all that, when it is completely characteristic of the spirit itself?

7. For example, this applies also to light. As you know, the earth is illuminated from the rays of the sun. Do you see in these rays the enlivening energy and the countless forms? No, you do not. You do not even sense the simple effective energy or power of the ray. And many people do not know any differently as they experience it every day, namely that this ray is not capable of any greater heat effect than the one it expresses daily.

8. But what would the observer say if these rays were concentrated or condensed into one focal point by a large magnifying glass and then expressed such energy that it could suddenly dissolve even the hardest diamond!? A layman would be surprised to witness the higher effects of these light rays, and he would probably say: Who would have believed that in those gentle sunrays that warm us every day lies hidden such an incomprehensibly fierce power!?

9. This applies as well to the ceremonial religious worship of the Saturnites! Even though the religious part appears to be completely spiritual, there is still in the religion of the Saturnites an extraordinarily strong "ether" which has, in the unification of its delightful light rays, a strong inner effect which is very difficult to understand for you people on earth. Therefore we can pose the question and say: Of what does this actual spiritual contained within the spiritual consist?

10. The actual spiritual contained within the spiritual which is already known to you lies within the living inner cognizance of the Great Spirit as well as all the levels and orderly relations and conditions which closely and affectionately connect the free human being with the heavens and the Great Spirit Himself and vice versa.

11. How do these human beings recognize the Great Spirit? First, they are given external instructions through words of the teacher about the nature of God. Then they are guided to the cognizance of the will of the Great Spirit and then to the practice of the recognized will.

12. These instructions are still the external religious worship, which is the path to the inner spiritual true life. It is actually already "spiritual wine" and the "non-concentrated light of the sun."

13. The active practice of the recognized will of the Great Spirit therefore resembles the "distillation" and "rectification" of a pharmacist and furthermore resembles the "concentration" of the sunrays by means of a large magnifying glass; or in other words, it is the actual independent active liberation of the actual prime spiritual being from its surrounding matter.

14. Are you now beginning to comprehend what the actual spiritual religious worship of the Saturnites consists of? I will give you the answer in a few words which will immediately open the door to the light. And these few words are: The rebirth of the spirit!

15. This rebirth must occur in the Saturnites as well as it must in you. Without it an inhabitant of Saturn will understand as little of the spirit and all the conditions and relations between him, the heavens and the Great Spirit as you do on earth.

16. It is absolutely unsatisfactory for the rebirth of the spirit that someone merely has the ability to see spirits, as is quite often the case on earth where many such people see these frightening and horrible apparitions of which they neither know nor comprehend any more than a blind person knows the colors of the rainbow. Clairvoyance is not as much a part of a complete or true rebirth as the determination of every action; that means that it is structured in such a way as it was established from the very beginning or prime origin in all its divine order and wisdom.

17. How and when will this take place? This can only take place when the spirit, through the strictest compliance of the prescribed path of God, rectifies himself from matter and finds himself again in a focal point and finally as a complete, whole existing being within himself, having stepped out of all sensuousness of matter as a complete unity in order to open up the new spiritual senses determinately and completely to the impressions and conditions of that world of which he himself is the actual original inhabitant.

18. Once a human being has attained this level, another activity begins to become effective in his whole being. Everything will begin to change his sight, his hearing, his feeling, his sentience. All his thoughts turn into forms which he will see; his will becomes an accomplished deed. His words become definite and unite with the thought and the will. Space is no longer a factor and time has shown him the last minute, because in a free spiritual existence you hear, see, feel, perceive, think, will, act and speak beyond time and space; that means, for such a person only one present exists, in which an

eternal past and an eternal future greet each other in a friendly manner. And something which is infinitely far away is no farther from his eyes than his own thoughts.

19. Every person on Saturn strives with all his might for this spiritual state of development; but he will not attain this state until he has completely fulfilled within himself all the conditions of life.

20. The control of one's own nature is not sufficient for complete fulfillment, but all the conditions of life that are known to him must become completely part of his personal proficiency.

21. When he has become a perfect master in all these conditions, he becomes free within himself and all his vital energy is born united out of all his sensual nature. When such a birth has taken place, then you already know what has happened to such a person; it is known as the "rebirth of the spirit."

22. Therefore, the faithfully continued practice of all conditions of life is what concerns the spiritual part of the religion of the Saturnites.

45

More Details Regarding The Path Of Rebirth And Becoming One With God

1. Do you now have an idea of what the spiritual religious worship of the Saturnites consists, and how it differentiates from the ceremonial worship?

2. Behind the spiritual ceremony lies well hidden the service of the spirit, which is constant, while the ceremony itself is successive in certain intervals of time.

3. Since this part of the religious worship of the people of Saturn is of the greatest importance for the people on earth, I will give you

another descriptive example through which you will be able to clearly differentiate these two kinds of religions.

4. Take, for instance, a student who wishes to embark on a career in the arts! Lets us assume that he wishes to reach the proficiency of a virtuoso in the art of music. If he would come to you, what would you say to him and what would be the first step? You would examine his abilities, and refer him to a reputable and orderly school and instruct him as to the conditions or give them to him in writing and tell him: "If you completely fulfill these conditions, you will without a doubt become a virtuoso. However, if you do not fulfill these conditions, you will be a strummer, but never a true virtuoso."

5. What will this student who is serious about his virtuosity now do? He will immediately make use of all his external diligence in connection with his inner volition and practice day after day, five, six or seven hours as it is required of him. He will also complete the school from A to Z and he will not practice anything other than those exercises which are necessary for the attainment of his purpose. When the student has practiced and developed in this manner for many years, he will be given by his teachers the encouraging certification that he has attained the level of an accomplished artist, since he now masters the instrument of his choice mechanically and spiritually at any degree of proficiency.

6. Now we have what we need! What was the prescribed practice or exercise for the attainment of the technical proficiency? It was nothing more than the well-ordered ceremonial part of his art cult or art worship. Did he practice day and night without interruption? Oh no! He only practiced the prescribed time necessary for the exercise.

7. How was the state of his endeavor and his will during the exercise? Was his will arranged periodically? Definitely not! The will

was without intermission similar to a good main spring of a watch, present in his spiritual and natural organism. This drive is the spiritual art cult of our music student through which he actually becomes completely whatever he has set his sights upon.

8. Once he has become an accomplished artist, what kind of a life will he live? Certainly he will not live the life of a student; instead he will live the life of a free master or maestro. Will he become, on account of that, an enemy of his former life as a student? Certainly not! He will continue as a great maestro to do what he did when he was a student, only now he will do it with truly great pleasure. He will diligently practice his scales and repeat all the other exercises that he practiced when he was a student. But with a great difference! What he did with great effort, ungracefully and with a great expenditure of energy as a student, he does now with the greatest of ease, without constraint, with confidence and full of inner spiritual significance.

9. As a student he played the scales, but he did not know the significance of what he played; now, as a master, he sees in the same scales countless new forms of which he had not the faintest idea before. That is how he, as a "reborn" master, practices the ceremonial art cult; but this cult is an entirely different kind of hearing, sight, feeling, perception, thinking and volition. And that is the spiritual *rectificatissimus* and it is the all-material ponderous and sensual dissolving combustion power of the rays of his spirit, and therefore by itself a purely spiritual cult.

10. If you transfer this to the actual life of a human being, whether of the earth, Saturn, Jupiter or the sun, everywhere and at all times there exists this twofold divine service which conducts itself as a path and as the goal or purpose for the path.

11. Whosoever continues to persevere on the path will reach his

destination. Once he has reached his destination, the path he has taken will never vanish from his memory nor from his remembrance forevermore. Not until he has reached his goal will he have a clear and complete view in his spirit over all the turns and relationships of the path he took.

12. You know the main goal the Saturnites are trying to obtain on their path is nothing but the Great Spirit Himself; they wish to become completely one with His will.

13. Once they have reached this goal through diligent practice, then the spiritual religious cult is complete; from this point on the actual impulse to reach the Great Spirit's will ceases. In its place steps the great unshakable longing to remain faithful forevermore.

14. And this exceedingly definite and firm wish is continuously the most inner divine service of the most complete spiritual kind of every reborn inhabitant of Saturn. This state can be reached on Saturn by everyone regardless of gender or age. This concludes the spiritual part of the religion of the Saturnites.

15. Since we have experienced in this manner everything that is of necessity and noteworthy in the mountains, we will shortly take a look at the lowlands and then take an excursion to the polar regions of this planet as well as the ring and the seven moons. That will be all for today!

46
The Inhabitants Of The Lowlands Their Construction And Housing And Their More Worldly Way Of Thinking

1. Concerning the depths or rather the lowlands of this planet, on different occasions we have learned many things about the lifestyle

of those who inhabit the lowlands on Saturn. But a few things remain to be said.

2. Nowhere on this planet will you find a city. However, there are regions such as the smaller lakes and the rivers where families live closer together than they do in the mountains, especially in the southern parts of some of the continental countries. But in these regions there are not as many family members living together in one dwelling as in the mountains. Here a family usually consists of the parents and their children. As far as the grandparents and great-grandparents are concerned, they usually live by themselves, but they also have some servants.

3. These people rarely live in the well-known dwelling tree (*gliuba*); their living quarters usually consist of a kind of tent-like structure built from trees, one tree placed over another in the following manner: The trees are positioned pyramid-like in a circular fashion, beginning from the ground to the apex, while lath or narrow strips of wood are nailed in circles around this dwelling. These laths are covered with all kinds of foliage, on the outside as well as on the inside. Towards the morning (east) is a spacious exit. And that is their residence.

4. As far as this type of dwelling is concerned, it cannot accommodate as many people as a dwelling tree in the mountains. However, such a tent house is still spacious enough to accommodate at least ten regiments of soldiers on earth.

5. The furnishings are similar to those which you find in the living dwelling trees in the mountains. Their sleeping accommodation consists of a frustrum pyramid-shaped back rest, where in front of the back rest there is a round elevation that serves as a table. And those are all the furnishings which this house contains.

6. Where do the people in the lowlands keep their tools, food

containers, clothing and food provisions? They are stored in the storage buildings, which are built in the same manner in the lowlands as they are in the mountains.

7. The temples are also built in the same manner as in the mountains, but they are often considerably smaller and not so sublimely magnificent as they are in the mountains. In the lowlands no single residence has its own temple; instead, several families have one temple.

8. That is all there is in regards to the construction and housing in the lowlands, except for some pleasure gardens, which are usually the property of a privileged patriarch. How these gardens are adorned and arranged was mentioned on the occasion when a description of the crustaceans was given, namely the snails. There are also snail shells in some of these gardens, which are prepared for the amusement of the children, which we learned about earlier.

9. The qualifications of a patriarch in the lowlands have also been discussed. Think back to the animal with the serpent's tail and fiery eye (*bauor*), and you will remember that whoever has a coat made from the skin of this animal can assume the position of the patriarch. It is not difficult to understand that this heroic kind of lowland patriarch is not held in as high esteem by the Great Spirit as a patriarch of the mountains. Furthermore, it is also understandable that the angel-spirits do not spend much time with the patriarchs who wear these coats. The reason for this is because the people living in the plains are much more worldly than those on the holy mountains. Even though altogether they are honest and upright brothers and sisters to each other, they often talk wisely and therefore consider themselves to be much more enlightened and intelligent than the inhabitants of the mountains. But when they begin to experience some bad times despite all their wisdom, they go on a

pilgrimage to the mountains, and you already know that they are instructed differently while they are there.

10. Therefore, you will find very few higher revelations in the lowlands and then only rarely, except for those who have come forth truly and faithfully from the schools of the mountains.

47

More About The Inhabitants Of The Lowlands
Commerce And Trade
Banishment Of Dishonest People
Mode Of Life, Matrimony, Procreation And Burial

1. In the lowlands, especially in the regions of the lakes and large rivers, the people carry on with some kind of bartering. They usually deal with the sort of merchandise with which women adorn themselves in the manner previously described.

2. It does happen that a merchant may think deliberately of somehow gaining an advantage over another person. But he is in real trouble when someone discovers the fraud. The women will scratch the merchant with all kinds of pointed objects. If he should be caught again after he has been given this kind of a lesson, then a ship will be prepared and the swindler and his family will be taken to a far-off place. He will remain at this place for a specified time, which is always considerable, or he may have to remain there forever, depending on the extent of his second fraud. On Saturn this punishment is known as purak or "eternal banishment." Whosoever is banished for a specific time only is allowed to take several "ship fruit seeds" (*chaiaba*) along, so that he can sow these seeds in the country of his exile and eventually prepare a ship from this fruit.

3. Those who are sentenced and exiled for life are not permitted to take the seeds of the *chaiaba* plant with them. It usually happens that the mountain dwellers of one or another country are informed by the spirits of the pitiful state of the exiled people. The mountain dwellers do not waste much time — they travel to where these people are exiled and welcome them and take them to the mountains, where they usually make the best human beings out of them.

4. Quite often they are given accommodation in the mountains to keep as their own property. It happens at times that the very people from the lowlands who sentenced these people to a life's exile may travel to the mountains and find shelter, hospitality and tutelage in true religion. When the guests recognize their hosts, usually one surprise replaces the other, whereby the guests cannot comprehend how their present hosts escaped the life-long exile and came to this place in the mountains.

5. On this occasion they show their amazed guests that the Great Spirit is capable of many things, of which the wisdom of the people who live in the regions of the lakes or ocean know very little. When the surprised questioners receive such an answer, they usually wring their hands and lament bitterly about the nonsense which is customary in the lowlands, upon which they are again seriously admonished that upon their return home they should contribute to the best of their ability to the elimination of many or most of such follies.

6. Through this approach it has happened in many of the large continental countries that the people of the plains are now on the same level as the mountain dwellers. But there are still countries where there exists a huge difference between the people of the lowlands and the people in the mountains.

7. Concerning manual skills and commerce, the lowlands and the highlands are comparatively similar, except for a few luxury trades

which are nowhere to be found on the mountains or highlands, where even the dyeing or coloring of a thread is considered a sin.

8. The nourishment (with the exception of the milk of the large cow) is the same in the lowlands as it is in the mountains. Only some of the patriarchal families have the beautiful large white rocks on the lakes prepared as pleasure castles (as mentioned in one of the previous chapters). While they occupy these pleasure castles they often tickle their palates with many delicacies that are a little more artificially prepared. But in due time they have to suffer for these indiscretions; and on account of this behavior, many wise medical doctors have much to do.

9. Remember, you have the same problem on earth! If you were to live a very simple life in accordance with nature and eat the fruits of the earth the way I prepare them — with only a few exceptions, and they simply have to be softened by cooking them on a fire — then your language would have four words less, one of which would be "doctor," and much less would it know of "medicine" or "apothecary" (pharmacy). It would not know the aforementioned three words because the first word, "sickness," would not exist. First you have very skillful cooks, followed by the medical doctors and the pharmacists with their even more skillful kitchens. And from these kitchens you obtain a completely perfect kind of fare (medicine) through which the ailment in the body of a patient becomes a permanent guest.

10. This is how it is in the lowlands of Saturn, though not to the same degree as on the earth. That is why the people of the lowlands do not grow nearly as old as the people in the highlands.

11. Concerning marriages, in the lowlands the patriarch joins husband and wife in marriage just as in the mountains, except that in the lowlands there is usually much more external pomp.

12. Procreation is the same as in the highlands, though it does happen in the lowlands that if the man does not possess enough belief and willpower, he undertakes a journey with his wife to the mountains in order to increase his willpower and belief. At this point, I do not have mention the purpose.

13. The burial of a human being in the lowlands is done in two different ways. For those who are on a higher spiritual level, the funeral is the same as in the highlands, whereas for the somewhat heathenish nations who regard the light ring as a kind of divinity, the funeral is considerably different. They place their departed on one of their ships and then sail with them onto the ocean, especially when the ocean is not far from their abode. When they have sailed far enough, the corpse is thrown into the water without any further ceremony; on this occasion they immediately find a living "grave," which very greedily eats the corpse. Once such a funeral has happened, the so-called grave diggers return and the minute they arrive in their homeland the entire funeral ceremony has come to its conclusion.

14. This is in addition to what you already know from the occasional information about the most remarkable and noteworthy customs or traditions of the valley and plains people on Saturn. We shall now leave the inhabited countries of Saturn and travel by the shortest route over the snow and ice regions of this planet to the ring. And this concludes today's information.

48
The Vast Northern And Southern Polar Regions
The Enormous Coldness Serves To
Temper The Destructive
Impulse Of The Primary Spirits Of This Planet
How Saturn Affects The Earth

1. It was mentioned in the beginning that this planet has actually only two climates, one of which is an altogether moderate climate where all the inhabitable countries are located and whose width, on the whole, occupies more than one-third of the polar region.

2. These lands of the moderate zone are surrounded by a continuous ocean in the north as well as in the south where you find only some *mud*-lands, but far away from the main continents the most northern or most southern parts are usually already frozen in permanent ice. This does not mean that one country in the south and the north is in such a location, but, should it be located in the northern hemisphere, then the most northern part is surrounded with permanent ice, and this is strictly for a completely natural reason, because on Saturn, as on earth, the polar regions belong to the permanent snow and ice.

3. What do these polar regions on Saturn look like? Even if you have an eye and a feeling for nature, no human fantasy or imagination will have the slightest idea.

4. The coldness of these polar regions is so severe that the cold of your own polar regions is like a well-heated oven by comparison. Not only does the water freeze there to a diamond, but at the lowest degree of temperature even the air freezes to ice rods which often protrude for miles into the ether. Through this immense pressure, the atmospheric air very often ignites in these regions, and that is

the reason why vast areas appear to be glowing for many miles, and this glow increases and eventually results in severe explosions.

5. Consequently these air ignitions, as a result of the immense coldness, decrease the electricity of the air; thus the coldness increases continuously and occurs in succession over a period of fifteen earth years. During the coldest stage, which lasts eight earth years, there are no ignitions, because the air has turned into a completely solid mass. After this period of time, the returning light of the sun gradually begins to dissolve the solid air, which after some dissolution begins to ignite again, and through these ignitions the explosions which are caused by them allow the dissolution of the frozen air to move closer and closer to the particular pole.

6. Are these dreadful regions inhabited? In a natural respect these regions are not inhabited by any living being, but are inhabited more so in a spiritual sense, because snow and ice usually provide the imprisonment of restless spirits. When the coldness is the most severe, then the peace spirits in these regions are the busiest. Because through this very act, which comes into being through the natural coldness, they are able to quiet down the fire spirits and temper their excessive fiery destructive force. When it becomes colder and colder in your regions on earth, you can always be certain that the destructive spirits on earth are being calmed and tempered by the peace spirits.

7. That these spirits are addicted to fire and destruction may be ascertained from the frequent air ignitions in the polar regions. The more these spirits want to enjoy themselves, the more persistently are they imprisoned by the peace spirits. The lower the temperature drops, the more intensive is the imprisonment of the fire-happy fiends, which, when imprisoned in this manner for several thousand years, finally submit and abandon their passion for fire. They appear

as a cloud filled with electricity, and ice-cold winds transform their electrical fire content into hail, which at times may cause some damage.

8. Who are these polar fire spirits on Saturn? They are not the spirits of departed human beings; instead, they are still primary spirits out of which the entire celestial body is actually formed, and they gradually pass over into a free existence in human form in accordance with the well calculated order of the Great Spirit.

9. At times it can happen that the spirit of a departed human being, when such a person has been very revengeful during his natural life, will be led back to his natural polar state. This happens very seldom on Saturn, but quite frequently on earth.

10. The difference between these two spirits is as follows: the spirits of departed human beings which are imprisoned in this manner will never return to a natural physical life, while this is always the case with the primary spirits, because they have to completely clothe themselves in this naturalness before they attain the ability to pass over to a free, independent and consequently absolute or separate life.

11. These primary spirits on Saturn are the ones which are of a highly destructive nature. This is why so many ancient prophets on earth mentioned that this planet would eat its own children. That is why these primary spirits, when they are no longer primary spirits, must first be thoroughly and properly prepared by the peace spirits before they can enter into an absolute and free life. If that were not done, no sun and no planet in the entire universe would be safe from their destructive powers.

12. This is the reason why Saturn is so far from the sun, so that the sun's rays are not capable of effecting such heat as they do on Jupiter, earth, Venus, and especially Mercury, whose inhabitants live

mostly in the polar regions and have to endure severe heat. Whereas on Saturn, as you know, the climate is moderate wherever it is inhabited, and even the moderate climate, if it should get too warm, is prevented by the permanent shadow of the ring from overheating.

13. Although the inhabitants of Saturn never enter these ice regions because their greatest fear is of snow and ice (this fear originates from their primary spiritual existence), they know exactly what it looks like in the polar regions, especially the enlightened mountain dwellers. Even the most enlightened have no great interest in viewing these regions or in the description of these regions. However, they have a much greater interest in the concept, description and viewing of the ring.

14. Why that is so will become clear to you when the ring is discussed in the next chapter. Concerning the polar regions, there is nothing more of importance which would be worthwhile to mention, except that those years on earth when the earth comes closer to Saturn are usually bad and unfruitful years. The reason for this is the excessive polar coldness of this celestial body, and this celestial body's effect often reaches, from a metaphysical point of view, several hundreds of millions of miles, similar to an invisible comet's tail.

15. In the vast solar system there are by far a greater number of negative comets, which completely consume all the sun's rays that fall upon them to such a degree that not even the smallest atom is reflected by them. The comets become visible when they have satiated themselves and begin to travel an orderly route. The comets are frequent guests of the planets and make themselves known at certain times for seconds in the form of shooting stars. However, concerning these negative comets, we shall learn more about them when we examine a sun. We only mention this subject so that you

may gather from this explanation how far-reaching and effective the polar coldness of Saturn is at times.

16. This concludes the actual information about this planet. In the next chapter we shall deal with Saturn's ring.

49

Saturn's Ring Has A Planetary Character Axis Rotation And Its Intended Purpose

1. Concerning the form, size and division of Saturn's ring, all of this was mentioned right at the beginning. What remains to be discussed is its usefulness and nature.

2. The ring forms a completely separate, compact and solid celestial body, and as far as the surface area is concerned it surpasses the surface area of the planet by many times. And since this area is larger, so is its physical volume many times greater than that of the planet itself.

3. Is the ring completely smooth or is it mountainous? Is there water on the ring and is the ring surrounded by atmospheric air?

4. This ring has all the components of a planet; namely, it has mountains which are extremely high, it has large lakes and rivers, and it is surrounded everywhere with atmospheric air. However, the water and the air are much lighter and finer than on the actual planet.

5. The ring also has a rotation of the axis around the common central point with the planet, but this rotation, as far as speed is concerned, is different from that of the planet. This has to be understood as follows: While the planet has almost turned around its axis two times, the inner ring, which actually consists of two rings which are connected with each other through many elliptical spheres, has barely rotated around its axis once. The middle ring has

an even slower rotation. The outer and largest ring requires almost seven Saturn days for one rotation.

6. You might ask: Why do these rings rotate at different speeds? Why do they not rotate at the same speed, and why not at the speed of the planet? Here you must consider the diameter of each ring, how one is larger than the other; knowing that must make it clear to you why each ring must rotate at a different speed.

7. For example, if the inner ring with its much greater diameter were to rotate at the same speed as the planet itself, then on account of the central point, centrifugal force, this speed would obviously tear the ring apart. If the second ring in its motion were to rotate at the same speed as the first ring or even the planet itself, then the second would be subject to the same fate, and that would apply even more so to the largest outer ring. The movement is exactly calculated so that no part of the ring will fall onto the planet, and this is controlled by the centrifugal force. The movement of every ring is therefore under the proper control, so that no part of a ring can be catapulted outwardly; that is accomplished through a properly controlled motion whereby the centrifugal force remains constantly in the proper ratio to the rings' specific gravitational force.

8. That is the well regulated control of nature of the ring which stands on the lowest potency. Now we shall deal with the second question:

9. What purpose does the ring serve, as far as the planet is concerned? Is the ring really, as some scientists have stated, only a caprice of the Creator or, an even worse scenario, merely nature's grandiose idea? Another one of these statements is that the Creator, when He created this planet, seriously ran out of ideas; that is why He let something which had magnificent beginnings stand unfinished and postponed the completion of the entire planet to a later date.

10. If one or the other scenario mentioned above is the case, the following explanations will show. In the last chapter you learned about the character of the primary spirits of this planet. If the ring would not permanently cause shadow that moderates the heat alternately, at times more northerly and at other times more southerly, where otherwise a hot tropical zone would develop by necessity, then the entire solar system would experience the might, power and force of the primary spirits of this planet and eventually this power would extend even to a shell globe.

11. Through this ring a perpetual moderate zone or climate is established in the habitable countries of this planet. The result of this is that the primary spirits of this planet do not ignite and therefore they no longer have the ability to devastate the celestial bodies in space.

12. That this is the case you can gather from the fact that the inhabitants of Saturn themselves have to be kept at all times in a condition of greatest respect and most conscientious obedience towards the Great Spirit. That is why not too much is lectured about love, only so much that they recognize love, but they are guided to have the greatest awe for love.

13. For this reason, love between husband and wife and the procreation of children is structured in such a way that the emotions of a human being are never stirred up too much. Everything is guided and settled with the greatest possible humility, which you were able to observe adequately when the description of the Saturnites was discussed.

50

More About The Various Purposes Of Saturn's Ring An Example Of Glass-Blowing And The Cooling Oven The Delicately Material Ring Inhabitants Their Way Of Life And Their Communication With The Inhabitants Of The Planet Absence Of The Animal World And Seedless Fruit Trees

1. You can see how well designed and calculated everything is, especially in regards to the ring which is located directly above the equator of Saturn! Furthermore, you have learned that on Saturn the whole of creation, beginning with the vegetable kingdom to the human being, manifests itself in giant bodies.

2. The question should arise: Is this all true, and if it is true, then why these giant-sized bodies? When the highest of all Spirits in the body of Christ while on earth had enough room in accordance with His order, then why do these Saturnian human spirits have such giant bodies?

3. The reason for this is that they do not experience any internal material or physical pressure. This pressure could ignite them when it originates from the outside and the pressure is exerted to the inside. For the same reason, they are also given extraordinarily delicate bodies, so that the easily excitable spirit would not find anything that could pressure him above his nature and easily stir him up to a complete ignition.

4. Now imagine these large, proportionately heavy bodies, which — as a result of the large volume of the planet itself and as a result of the fact that the gravitational or attractive force in proportion to the planet — should be much heavier than proportionately similar bodies on your much smaller earth. Would anything be gained for

these kinds of spirits through the earlier careful undertakings, if they were to be left in their bodies with such great ponderousness?

5. That is why My science, which is on a somewhat higher level than that of your scientists, invented this ring, through which the gravitational force of the planet is considerably decreased, so that these giant bodies in proportion to their planet are almost a hundred times lighter than your bodies in proportion to your much smaller earth.

6. This is an extraordinary, very important purpose of this ring which has not been mentioned before. Even if it appears small by comparison through a telescope, this ring is of the greatest importance — because it is not only a ring around a planet, but should be considered a strong band over the entire universe.

7. Now the question arises: Is this the only and final purpose of this ring? Oh no! We shall learn about another purpose, which is by far greater and more important than the earlier ones. But before we shall touch upon and examine the main purpose, we must ask the question: Is this ring inhabited?

8. That is why I say here: If the main purpose arises from that, then the ring must be inhabited. But by whom and how, that is another question.

9. But before I answer this question, as an example, I would like to direct your attention to one of the trades which you practice on earth, and that is the manufacture of glass.

10. Once you have ground up the ingredients necessary for the manufacture of glass and mixed it with the necessary salts, then these ingredients are placed into a melting pot, and by applying the proper amount of heat these ingredients become liquid. Observe this molten white-hot glass matter! That is the condition of the Saturnian human being in his body on the planet.

11. What happens to this substance when it reaches the proper liquidity? Probably all kinds of different vessels, containers, jars and so on are made from this substance in the well-known manner, namely by means of the air which fills the lungs of the glass blower and employing a glass blowers' pipe. Here we have our Saturnite as a fine, translucent spiritual human being who, through the rebirth, relinquishes his physical body. Through this the spirit already assumes a solid form.

12. When the glass vessel is produced in this manner, the blower's pipe is removed and then the vessel is placed into another pot and into a cooling oven. Now we have arrived at our ring, because when a human being dies on Saturn, he is, in a manner of speaking, removed from the blower's pipe of the great Glass Manufacturer and is taken in another pot into the cooling oven. This cooling oven is the ring!

13. The first ring is for the cooling of the greatest heat. The second ring is for a further cooling off. And the last ring is for the pliancy, and only after that is every liberated human spirit on this planet capable of accepting love.

14. I am of the opinion that, on account of this example, any further explanation is unnecessary, because the explanation is so clear. However, that does not mean that some would not ask the question: What is the purpose of giving these spirits a material place to stay?

15. The answer to this question is very easy, because the spirits of the Saturnian human beings, when they step out of their first body, are not immediately pure spirits, as you can gather from their easy reappearance after death and the quick decomposition of their bodies. When these spirits enter the great ring, they still have a kind of material body, which of course is much lighter, more delicate and

more pure than the earlier one they had while on the planet. And even this body becomes continuously purer and more spiritual, as the Saturnian human being passes on to the higher spheres of the ring.

16. The inhabitants eat, drink and live on the rings just as they did on the planet, with only one difference: all the products are proportionately finer and more subtle, just like the human beings who are present here.

17. The difference between this ring and the planet is that upon this second world there are no animals present; however, there are fruit trees, but none of them have any seeds for propagation, for they grow from the soil like mushrooms on earth.

18. That these ring inhabitants can proceed from the ring for a short time to the planet itself upon the inner wish of the inhabitants of the planet, and with spiritual speed, can be ascertained from many such appearances of these spirits among the inhabitants of the planet itself.

19. But since the dwellings and the living conditions of the spirit human beings on the ring are incomparably more magnificent, sublime and comfortable, the ring spirits are not inclined to remain any longer on the planet than the will of the Great Spirit finds it necessary. That is why they are always pleased when they can return to the ring again.

20. Now you know everything that is necessary and noteworthy about the ring. In the next chapter we shall deal with the moons of this celestial body and conclude the communication about this celestial body.

51

The Saturn Moons Without Any Axis Rotation
Natural Life Exists Only On
The Opposite Side Of The Moon
The Moons As Post-Schools Of Saturn
And Pre-Schools Of The Ring And
The Purely Spiritual Life
Closing Words

1. Regarding the moons, they have the same relationship to their planet as does our moon to the earth, only there is a gradation through the moons, which of course cannot exist on earth because the earth has only one moon.

2. As far as the movement of the moons is concerned, they always turn the same side half towards their planet.

3. For this reason their habitableness is twofold, namely a spiritual and a natural one. Therefore the opposite side of every moon facing the planet is inhabited by human beings and animals, and has vegetation, water and air and everything that is necessary to support natural life.

4. Human beings that live naturally on the back of the moon are of course considerably smaller than those on the planet and on the smaller moons; they do not even have the size of the people on earth. On the larger, outermost three moons the people are taller than on earth.

5. The natural human beings on the moons are in constant spiritual contact with the actual Saturnian human beings, so that the spirits of those Saturnian human beings who did not prepare themselves during their natural life span to reach one of the rings must therefore first endure the spiritual life of one or more moons in

accordance with their level of development before they can be accepted on the lowest ring.

6. What do these human beings do on the moons? And what kind of Saturnian spirits come to a moon? Spirits that come to a moon are selfish and heathenish, those that worshiped and revered the ring as a deity throughout their Saturn life. On every moon they first appear on the natural side and, through their bodies, they view that which is natural and live there as natural human beings, but they cannot see the ring which formerly was their idol.

7. When, through this, they have been weaned off the ring and have also lost their planet, then they move to the side of the moon facing the planet, from whence they can see the planet and the ring as an almost concrete body. Through this it becomes gradually clear to them, while they are being instructed by other higher spirits that come to them, that the ring is not at all some kind of deity or the residence of a deity, nor is it the pathway upon which the Great Spirit walks above the heavens. But they can see with their own eyes that it is only a materially solid celestial body which is placed around the actual planet. And this ring has been created by the Great Spirit so that the spirits of the departed human beings on that planet are prepared on the ring for a higher life of which they thus far have no idea.

8. When these spirits have learned this, through teachings as well as through their own intuition, they soon completely abandon their false beliefs and make inquiries as to the abode of the Great Spirit. It is pointed out to them that they will receive this information on the rings. They have completely approached the purely spiritual state and finally also have completely passed over into that state. After this they experience a longing for the ring, but yearn more so for a purely spiritual state; when that happens they are immediately transferred to the ring.

9. Now you have the answers to the previous questions, but in time this question might arise: Why are seven moons required for this purpose? Could not one moon fulfill such a simple task?

10. That would be true if they were spirits of a different nature; then one moon would suffice. However, the Saturnian spirits have their seat under one knee in the Great Man of Creation or the Great Cosmic Man,[39] and that is more than enough of a reason. Because the legs are the outermost foundation of life and on the leg itself again the joints. If, for example, the body sustains an injury to an arm or to the skin or anywhere on the body, the body can still stand upright and move around to look for help. If, however, one of the legs, especially one of the joints, sustains a severe injury, then the whole body is inhibited, and the person will probably fall to the ground unable to move and therefore unable to seek help. And that is the reason why the legs of every human being are built stronger than all the other parts of the body.

11. The inhabitants of Saturn represent the most important part of the leg under the knee of the Great Man in Creation (also called the Great Cosmic Man), and you have already heard of this great spirit being on other occasions. For this reason great attention must be paid regarding the spirits of the Saturnian human beings, and that applies to every single spirit as to which one of the seven primary spirits[40] (out of which every spirit consists) is in the most

---

[39] The Great Cosmic Man or Great Man of Creation means the great transformed primary spirit of Lucifer. For further details see *Robert Blum – From Hell to Heaven*, and *The Fundamental Principles of Life*, by Jakob Lorber. – ED.

[40] The seven primary spirits of God are: (1) Love, (2) Wisdom, (3) The Effective Will of God, (4) The Order, (5) Divine Seriousness, (6) Patience, and (7) Mercy. Each of these spirits originates from the previous one. For more information refer to *The Great Gospel of John* and *The Fundamental Principles of Life* by Jakob Lorber. – ED.

dangerous position. And this is the purpose for the existence of the seven moons, so that in one or the other moon one or the other of the seven primary spirits can be calmed down and brought into the appropriate order with the remaining six primary spirits. From this explanation you will be able to gather why this planet was assigned seven moons.

12. Now you know everything there is to know regarding these moons. Their distances from the planet and their size was mentioned at the beginning of this book. Therefore, this concludes the descriptions of the moons.

13. Since we have learned about the planet, the ring as well as the moons, we shall conclude this work.

14. Some of you might still ask: What benefit or purpose does this whole message about Saturn serve? I will answer as follows:

15. First, each of you who has read this message should take this example as a serious reminder of how completely different the inhabitants of this planet respect My will than do the inhabitants on earth.

16. Secondly, you should gather from the entire message how My love, wisdom, might and Fatherly solicitude reaches considerably farther than the arrogant human intellect's foolish sense can conceive.

17. Thirdly, this observation should guide people on earth to the deepest humility, from which a human being should see who he is and Who I am, their God, Creator and Father.

18. He should look into his heart and contemplate on the greatest blessing and compassion that has been bestowed upon him, and that I, the only Lord and Creator of these miraculous works, consented to choose the earth, this dirty little planet, as the birthplace of My infinite love, compassion and mercy, and consequently all the fullness of My divine being.

19. For this reason I will allow you to inspect the sun[41] as well as a few other planets. It will not be described in so much detail; although brief, it will be sufficient.

20. And with this promise I conclude this communication. My blessings, My love, mercy and compassion be with this promise! Amen!

---

[41] *The Natural Sun* by Jakob Lorber. – ED.

Titles By
Merkur Publishing, Inc.

The New Revelation

A Spiritual View of Life

The Childhood of Jesus

Fundamental Principles of Life (3 volumes)

Gems from the Master of all Masters

The Great Gospel of John (6 volumes)

Jesus' Words

The Lord's Sermons

Sunsets into Sunrises

Hermetics

Frabato the Magician (introduction)

Initiation into Hermetics (Vol. I)

The Practice of Magical Evocation (Vol. II)

The Key to the True Kabbalah (Vol. III)

Franz Bardon: Questions & Answers

Alchemy Unveiled

Seven Hermetic Letters

Prophecy

Through a Window Small

Philosophia Mystica: The Prophecies of the Prophet Daniel

The Advent of Christ

Creation

Earth & Moon

Saturn

Health

The Lord's Book of Life & Health

The Healing Power of Sunlight

Metaphysics

Magic: Principles of Higher Knowledge

Up-coming Titles

The Natural Sun

The Spiritual Sun (Vol. I and II)

The Mysteries of Life

The Mysteries of Creation

Merkur Publishing, Inc.
PO Box 171306
Salt Lake City, UT
U.S.A. 84117
Toll free (800) 204-2473